518

D1764936

Network Modeling, Simulation, and Analysis

Network Modeling, Simulation, and Analysis

edited by

Ricardo F. Garzia
Mario R. Garzia

AT&T Bell Laboratories
Columbus, Ohio

Marcel Dekker, Inc.   ■   New York and Basel

Library of Congress Cataloging-in-Publication Data

Network modeling, simulation, and analysis / edited by Ricardo F.
  Garzia, Mario R. Garzia.
      p.   cm. -- (Electrical engineering and electronics ; 61)
    Includes bibliographical references.
    ISBN 0-8247-7876-6 (alk. paper)
      1. Computer networks--Mathematical models.   2. Computer networks-
  -Simulation methods.   I. Garzia, Ricardo F.   II. Garzia,
  Mario R.   III. Series.
  TK5105.5.N46623   1990
  004.6--dc20   90-30826
                                                                CIP

This book is printed on acid-free paper.

MARCEL DEKKER, INC.
270 Madison Avenue, New York, New York 10016

Current printing (last digit) :
10 9 8 7 6 5 4 3 2 1

PRINTED IN THE UNITED STATES OF AMERICA

Preface

This book was written for the purpose of compiling, in a single volume, a series of topics falling within the general area of network modeling, simulation, and analysis for both circuit-switched and data networks. The book also serves the purpose of presenting some recent work in the areas of compartmental models for the analysis of communication networks; closed-form solutions for the evaluation of network performance measures, adaptive routing and design for reliable distributed networks, and mixed voice/data networks. Although the book deals primarily with circuit-switched and data networks, each possessing somewhat different modeling and analysis requirements under the common thread of network applications, most of the techniques discussed here have applications in other areas of science and engineering, for example, biological and manufacturing systems.

Portions of the material covered were presented at an IEEE workshop held during the Seventeenth Annual Simulation Week, Tampa, March 1984. This turned out to be a successful workshop which met with a large amount of interest and enthusiasm from the workshop's participants. The material covered there, and in the book, had several different origins. The sections on local area networks arose from lecture notes that were used in the presentation of mathematical modeling courses and seminars both within and outside AT&T Bell Laboratories. The work dealing with the design and performance of reliable distributed networks and the modeling of network dynamics is the result of (ongoing) investigations performed at AT&T Bell Laboratories dealing with robustness and vulnerability issues of circuit-switched telecommunications networks. The work on mixed voice/data networks consists of Ph.D. research carried out at Texas A&M University. All the work presented is of current interest and most of it does not appear (to our knowledge) elsewhere in book form. Some of the results, however, appear in recent issues of journals and conference proceedings.

The book's title has been chosen to reflect the breadth of topics covered and does not mean to imply an in-depth coverage of each of these topics. In developing the chapters in this book, we have aimed at presenting a methodology for developing network models, both analytical and computer simulations, that can be used to study different types and aspects of networks. A second aim has been to present some recent results on network modeling. Through the incorporation of examples drawn from actual case studies, the book investigates techniques for analyzing these networks and presents results that are of current interest. The book should be of value to researchers and analysts working in the area of network modeling and simulation as well as for graduate students in the computer science or engineering disciplines.

The book is composed of four parts: Part I, Modeling and Simulation; Part II, Computer Networks; Part III, Wide Area Communications Networks; and Part IV, Mixed Voice/Data Networks. The book has been organized in an orderly and modular fashion. After a review of background material in Part I, the reader may proceed directly to the parts of the book that are of particular interest.

Part I has been included with the intent of making the book somewhat self-contained. In Chapter 1, we present some concepts associated with modeling in general, and with mathematical modeling and computer simulation in particular. Discrete event simulation and its relationship to mathematical modeling is briefly discussed. Network modeling is emphasized, and the general methodology for this type of application is introduced. Chapter 2 presents background material on random variables, and Chapter 3 presents necessary background on compartmental models.

Part II covers model development for computer network systems. The model development methodology is centered around the general equations for queuing network systems in equilibrium and the generating function of the network. Chapter 4 contains an introduction to birth-death processes. Both continuous time and discrete time processes are discussed along with the general equilibrium solution for such systems. Chapter 5 introduces birth-death queuing systems, with particular emphasis on M/M/1 and M/M/∞ systems, and Markovian queues in equilibrium. Chapter 6 concentrates on queuing network models that utilize the generating function of the network. This chapter follows the approach developed by J. P. Buzen concerning the analysis of queuing networks with a "product form" solution. Calculation of network performance measures is discussed, and iterative and closed-form solutions for obtaining these measures are given by making use of the generating function of the network. Applications to open networks are presented and used to evaluate computer network performance for networks operating under load peaks. Model development for a wide set of computer network architectures is considered. These computer networks are then analyzed in detail. In this chapter we also present some extensions of the previous mathematical development necessary for large computer networks along with some applications.

Part III covers issues related to the performance and reliability of distributed networks, with particular emphasis on network management controls, adaptive routing, and network design. This part presents actual case studies as well as a discussion of the network model and computer simulation employed in these studies. The results are recent and of general interest in network analysis and design. The main emphasis of these studies will be on maximizing an objective (e.g., network-carried load) subject to traffic overload as well as node and link outages. In Chapter 7, a discussion of network topology and network reliability issues for distributed communications networks is presented. A study of adaptive routing schemes for nonhierarchical communications networks is presented in Chapter 8. The adaptive routing schemes investigated include both local and global adaptive routing. Chapter 9 presents analytical models for the analysis of communication network dynamics. The models described are in terms of global variables (e.g., carried load, blocking probability) and are aimed at analyzing the steady state and transient behavior of networks. This approach leads to a qualitative understanding of network dynamics. The models employed belong to the class of compartmental models that, although seldom used for communication networks, have found extensive use in the chemical and biological sciences.

Part IV of the book is concerned with the modeling of mixed voice/data networks. Recent advances in technology have shown the feasibility of transmitting voice and data simultaneously on a common transmission medium. The ultimate objectives of integrating voice and data on a common carrier include the economics of equipment commonality; increased utilization of resources; and a unified approach to network operations, maintenance, and administrative policies. Chapter 10-12 address the design and analysis of such integrated networks. In Chapter 10, we present a brief background of the integration concept and develop the structure upon which a simulation program has been written. Chapter 11 develops the procedures that can be used for integrated network performance analysis using the simulation tool described in Chapter 10. A methodology for optimizing the design of an integrated computer network is presented and also demonstrated in Chapter 12. The emphasis in these three chapters is on the procedural aspects of modeling and analyzing integrated computer networks rather than on the specific tools that have been generated to implement them.

We wish to thank our colleagues Professor J. Dshalalow, from the University of Pittsburgh at Bradford, B. D. Huang and N. D. Prabhakar, from AT&T Bell Laboratories, and the members of the AT&T Bell Laboratories Book Review Board for their many insightful comments during the review of this manuscript. Professor P. H. Schmidt, from the University of Akron, contributed the material on the direct calculation of G coefficients for the generating function of the network, Chapter 6. We also express our gratitude to Marcel Dekker for proposing and collaborating in the process of preparing this book and to AT&T Bell Laboratories for their support in its preparation.

Grateful acknowledgement is made to the following sources for permission to reproduce copyrighted material (omissions brought to our attention will be corrected in any future edition): Figure 2.10, L. Kleinrock, *Queueing Systems Volume I: Theory,* John Wiley & Sons, 1975; Figure 6.12, M. Ajmone Marsan, B. Balbo, and G. Conte, *Performance Models of Multiprocessor Systems,* MIT Press, 1986; Figure 12.4, M. Gerla and L. Kleinrock, "On the Topological Design of Distributed Computer Networks," *IEEE Transactions on Communications* COM-25 (1), pp. 48–60, January 1977.

Ricardo F. Garzia
Mario R. Garzia

Contents

Contributors

Mario R. Garzia   AT&T Bell Laboratories, Columbus, Ohio

Ricardo F. Garzia   AT&T Bell Laboratories, Columbus, Ohio

Mark J. Kiemele   Department of Mathematical Sciences, U.S. Air Force Academy, Colorado Springs, Colorado

Clayton M. Lockhart   AT&T Bell Laboratories, Holmdel, New Jersey

Network Modeling, Simulation, and Analysis

I
Modeling and Simulation

1

Introduction to Modeling and Simulation

Ricardo F. Garzia and Mario R. Garzia

AT&T Bell Laboratories
Columbus, Ohio

1. INTRODUCTION

In the study of systems one is often led to modeling and simulation as the most suitable approach for investigating properties and behavior. The system can be either naturally occurring, manmade, or abstract. Often, the importance of modeling and simulation in the study of such systems results from the

- impossibility of dealing directly with the system (e.g., the system may not yet exist).
- cost of studying the system directly, which may be too high (e.g., physiological systems).
- impracticality of dealing directly with the system (e.g., time-consuming).

Modeling the system consists of abstracting its features and properties paying particular attention to those that are of interest to the study. This is usually not just a reasonable approach, but a necessary approach to obtain a manageable model. That is, a model that can be properly analyzed to yield answers to questions posed about the system. It is often the case that all the details about the system are not fully known, making this the only possible approach.

1

The term *Simulation* has been defined in various ways over the years. For the purposes of this discussion, simulation is (loosely) considered to be the realization of a model, used to represent a system, in a form suitable for deriving information about the system's characteristics and properties of interest. The term *suitable form* means that we are able to compute (derive) qualitative and/or quantitative information which elucidates certain aspects of the system, usually by means of a computer program. Thus, one must first have a model of the system before a simulation can be developed. The system model itself can range from a basic conceptual model to a precise set of mathematical relationships describing the system's behavior. Based on the chosen representation, the associated model can be deterministic or stochastic in nature. Within each of these subdivisions models can be further categorized in terms of the way in which their state changes as a function of time. The associated simulations can therefore be classified into one of three types: *continuous time, discrete time, or discrete event*. Continuous time simulations arise from models described by ordinary differential or partial differential equations. Discrete time simulations arise from models that can be expressed as difference equations. The third type, *discrete event simulation*, arises from systems that can be modeled as a sequence of countable events, where it can be assumed that nothing of interest takes place between those events. Queueing and communications networks are examples of such systems. Discrete event systems cannot be modeled by standard difference equations since the evolution of the system is not simply controlled by the passage of time but by the interaction of events in time. The relationship between the system, model, and simulation is depicted in Fig. 1.1.

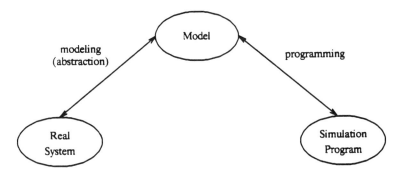

Figure 1.1   Relationship Among the Real System, Its Model, and Simulation.

The modeling step shown in the figure is somewhat different for discrete event systems than for continuous or discrete time systems. This difference can be attributed to their historical development. Continuous (and discrete) time simulation arose from a need to numerically solve systems of differential (difference) equations for which a closed-form solution was either impossible or impractical. Nonetheless, over the years a large amount of theory was developed on the qualitative and quantitative behavior of such systems and the associated set of equations used to represent them. Aside from gaining an understanding of such systems, this theoretical background also helped to develop appropriate simulation languages and to supply an understanding of how such languages operate. These simulation languages need only be concerned with the efficient and accurate solution of the corresponding system of equations.

Such a theoretical background was and still is largely non-existent for discrete event modeled systems. For these types of systems, practice has preceded theory. Simulation languages have been developed that can be used to represent the behavior of the discrete event system. A general theory for modeling such systems is not yet available. Among other things, the lack of such a global theory of discrete event dynamical systems causes problems in the understanding of and belief in discrete event simulation. As a result, discrete event simulation is often considered the tool of last choice. Without such a theory it is hard to categorize and compare the different simulation languages currently in existence, since in this case there is no "set of equations" with which we can model such systems and whose efficient solution would lead to an appropriate simulation language. Recently, steps in this direction have been taken by various researchers in the field.

Owing to a lack of theoretical guidance, different languages have, historically, approached discrete event simulation in somewhat different ways. At times, the way in which a given language approaches the problem has been perceived by some as a *theory* of discrete event systems, completely confusing and obscuring the fundamental properties of these systems. In using a specific simulation language (e.g., GPSS, SIMSCRIPT) one must therefore split the modeling step, shown in Fig. 1.1, into two separate steps, one of which directly depends on the specific simulation language chosen.

This is represented in Fig. 1.2, where it is seen that an extra step between the modeling and simulation steps is included; we call it the language specific model. This step consists of representing the derived model in terms of a model that is appropriate for coding in the chosen

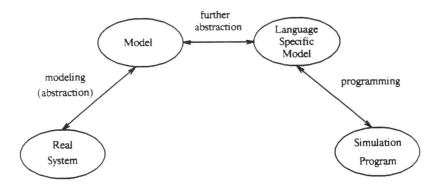

Figure 1.2   The Complete Modeling Process.

simulation language. For example in GPSS, this would consist of defining
the model in terms of facilities, storages, and transactions. In SIMSCRIPT
it would require deriving a model in terms of processes and resources.

2.   MODELING AND SIMULATION METHODOLOGY

Language specific modeling tends to highlight differences in modeling
methodologies. However, viewed from a system theoretic perspective,
there exists much more significant commonality than might be realized on
first glance. That is, by defining the system in terms of objects and relations among these objects important fundamental considerations are
revealed. In two books, *The Theory of Modelling and Simulation* [6] and
Multifaceted Modelling and Discrete Event Simulation [7], B. P. Zeigler
lays down a foundation for a modeling methodology with special emphasis
on modeling as applied to discrete event simulation. As described in this
work, five basic elements can be associated with the modeling process;
these are the *System*, *Base Model*, *Experimental Frame*, *Lumped Model*,
and *Computation*. A (lumped) model of the system is valid only under certain conditions (as specified by the experimental frame), whereas the base
model is a valid representation of the system under all conditions (all
experimental frames). Although the base model is not fully attainable
because of its high degree of complexity and the fact that we do not necessarily know everything about the operation of the real system, it is nonthe-
less used to derive (many times by lumping together elements of the base

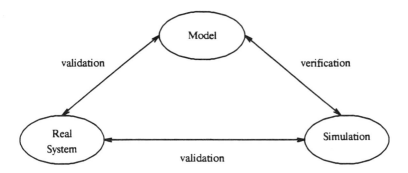

Figure 1.3   Validation and Verification.

model) models by assuming away some of the base model's complexities. Validating the model under the specified experimental frame guarantees that the reduced complexity is appropriate for the study in question. Alternative system theoretic modeling methodologies have been proposed.

When developing a model of a system, it is necessary to validate the assumptions made regarding the real system characteristics under consideration. Once the simulation (computer code) is developed, it is necessary to verify that the simulation indeed does what the model intends for it to do (no program bugs). This is the verification stage. Finally, once the simulation has been verified, the final validation step consists of comparing the model's (simulation) input/output behavior with that of the real system. The necessary validation/verification steps are shown in Fig. 1.3.

Validation itself can have different meanings. For example, we can consider a model as having replicative validity if it is capable of reproducing known system input/output behavior. The model may possess predictive validity if it can be used to predict system input/output behavior (i.e., behavior not previously observed). Finally we can say that a model has structural validity if its components directly relate to (and interact as) the corresponding system components. The validation steps alluded to in the figure, therefore, are directly dependent on the type of model validity sought. Clearly one must properly specify, at the outset of a study, the study's objectives and hence the conditions (experimental frame) under which the model will be valid. This will not only assure that the right model is developed and that a clear definition of the study's objectives is available, but also that the model's results will be properly interpreted, i.e., within the confines of the experimental frame for which the model is valid.

Of course if, during a study, the study's objectives change, it will be necessary to change the model and conditions under which it is valid.

The preceding discussion makes a clear separation between the model and the computer program (computation process). This distinction is not always made in practice, where computer code is often confused with the model. Stressing this distinction leads to an understanding for the need of a validation step to assure that the model is a good realization of the system within the specified conditions, and a separate verification step to assure that the model computation is faithful to the model's instructions.

Once a model has been developed, it is necessary to communicate it to others. A model description should include a specification of the components making up the model, the variables used to describe each of these components, parameters used in the model, and a description of the interactions between the model components. Conditions under which the model is valid are usually presented as a set of assumptions about the system. More formal ways of describing a model entail a full description of the time base (e.g., discrete or continuous), the input, output, and state sets, and the mathematical equations describing the model interactions and output. Developing an appropriate model description is essential for an understanding of the model and for continued model use.

In practical situations one is often faced with multifaceted studies. Building a single model to answer all the questions raised in a given study entails the development of a model whose size and complexity approximate that of the corresponding base model. Even if possible, such a model would not be practical since it would require long computation times and therefore would make it impossible to complete all necessary studies with the needed detail. This is a common pitfall in the development of discrete event simulations, in which one model (and associated computer simulation) is built to answer all existing questions about the system, and perhaps even future questions that may arise. Such projects usually result in large computer simulation programs, filled with details that may not be necessary, requiring large (perhaps even prohibitive) simulation run times. The attempt to develop such multi-purpose models and simulations is simplified because, in contrast to analytical models, basically any level of detail can be incorporated. Such studies make discrete event simulation appear as an unacceptably expensive and time-consuming endeavor, adding to the perception by many of simulation as the tool of last resort.

If we set aside the idea of a single model, multifaceted modeling methodology [7] suggests developing different models corresponding to different views of the system (different experimental frames). The selection of appropriate system views is important to make sure that all system characteristics of interest are properly represented. A drawback to having separate models, however, is that there is no single, multipurpose model with which one can study global system issues, as would be possible with a single large-scale model (i.e., the base model). To be able to study the multifaceted aspects of the system it is necessary to develop a hierarchy of models (or equivalently, a hierarchy of experimental frames). Models high on the hierarchy pay little attention to system details, whereas models lower on the hierarchy are detailed models whose objectives are to obtain information about a portion of the overall system.

Once a system of interest, study objectives, and corresponding system model have been specified (the latter step properly validated), the next step is to develop a language specific model and corresponding simulation program. When dealing with an appropriate class of systems, problem oriented simulation languages (e.g., SIMFACTORY by CACI Inc.) make these latter two steps easier than general purpose simulation languages. Since these special languages are specifically developed for a particular type of system, they are able to reduce the effort needed to develop a language specific model, and even the amount of coding effort required. General purpose languages, on the other hand, can be used to model general classes of discrete event systems. When contemplating the use of one of these languages for a specific application, it is necessary to select the language that is most appropriate for the system/model under consideration. That is, the language which facilitates the step from a system model to a language specific model. Along these lines, one can classify discrete event simulation languages according to their simulation strategies as shown in Table 1.1. Each of the currently available languages focuses on a specific strategy; however, as can be seen from the above list, several of them merge strategies and/or offer the capability of using different views. These *world views* show up in model development as well as the way in

Table 1.1 Simulation Language Strategies

Strategy	Simulation Language
event scheduling	SIMSCRIPT
activity scanning	GPSS
process interaction	SIMSCRIPT, GPSS, SIMULA

which the simulation language is implemented; for example, SIMSCRIPT - process/resource, GPSS - transaction/facility, SIMULA - process.

3. A VIEW OF SIMULATION

In the 1950's, DC Analog Computers started being used, in an effective way, to solve a set of differential equations. From our point of view, this was the practical starting point for simulation. Let us consider for a moment this development. To solve ordinary differential equations using DC analog computers, we need to make an analogy between the variables in the problem, and the machine variables in the computer. In other words, each variable $(x_1, x_2, \ldots, x_n)$ in our problem will be represented by a voltage $(v_1, v_2, \ldots, v_n)$ in the analog computer, refer to Fig. 1.4. The voltages in the computer will vary with time as prescribed by the differential equations. This resulted in the name *analog* for this type of machine. A discussion of this subject appears in [4].

This was really the simulation of a set of differential equations using electric circuit properties. Although the name *simulation* was not used at that time, it was implicit in this process. The term used back then was *Analog* which emphasizes the analogy between problem variables and computer variables.

After that time, digital computers became more popular and we began thinking about effective simulation languages to implement the solutions of the set of differential equations. The CSMP- IBM 1130 version was a

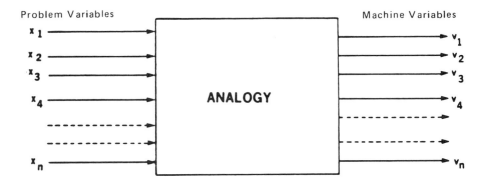

Figure 1.4   Problem Variables, Machine Variables.

popular tool for these purposes in the 1960's and the name *Simulation* became popular.

We now present examples of the various types of simulations described earlier. Although there exist mixed continuous/discrete simulations, we choose here to concentrate on the three basic types.

3.1 Continuous Simulation

Continuous Simulation is the simulation of a system of linear or non-linear differential equations. This set of differential equations can be represented in the *time domain* or *transform domain*. By the latter we mean the Laplace transform domain. Model development may proceed as follows:

- Write the set of integro-differential equations (linear and non-linear) which represent the process to be simulated. In doing this task you need to have a good understanding of the laws governing the process under study. Therefore, understanding basic concepts such as Newton's law, Kirchoff's laws, Maxwell's equations, network topology, state variable methods, laws of thermodynamics, chemical dynamics, fluid mechanics, etc., becomes mandatory.
- The set of integro-differential equations is transformed into a set of differential equations. This is important for the following step.
- The highest derivative in each differential equation is placed on the left-hand side of the equation and the rest on the right-hand side.
- The highest derivative is assumed to be known. Therefore, by successive integrations this derivative becomes the problem variable. This process is repetitive for each differential equation. Also, in all required integrations, we include the corresponding initial conditions.
- It is possible to obtain the highest derivative by algebraic manipulations.

We see therefore that, in continuous simulation, the *heart* of the process is integration.

Example 1.1 Suppose that we have the mechanical system shown in Fig. 1.5. We are interested in the behavior of $x_1(t)$ and $x_2(t)$ given the initial conditions

$$x_1(0) = 0.175 \quad \dot{x}_1(0) = 0.0$$

$$x_2(0) = -0.167 \quad \dot{x}_2(0) = 0.0$$

For the mass m_1 we can write

$$m_1 \frac{d^2 x_1}{dt^2} + f_1 \frac{dx_1}{dt} + k_1 x_1 + k_2(x_2 - x_1) = 0 \qquad (1.1)$$

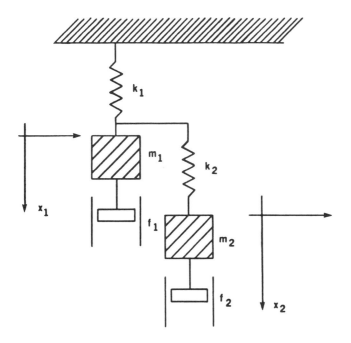

Figure 1.5   Mechanical System.

or,

$$\frac{d^2x_1}{dt^2} + \frac{f_1}{m_1}\frac{dx_1}{dt} + \frac{k_1}{m_1}x_1 + \frac{k_2}{m_1}(x_2 - x_1) = 0 \qquad (1.2)$$

Similarly, for the mass m_2

$$\frac{d^2x_2}{dt^2} + \frac{f_2}{m_2}\frac{dx_2}{dt} + k_2(x_2 - x_1) = 0 \qquad (1.3)$$

Solving Equations (1.2) and (1.3) for the term involving the highest derivative, we have

$$\frac{d^2x_1}{dt^2} = -\frac{f_1}{m_1}\frac{dx_1}{dt} - \frac{k_1}{m_1}x_1 - \frac{k_2}{m_1}(x_2 - x_1)$$

$$\frac{d^2x_2}{dt^2} = -\frac{f_2}{m_2}\frac{dx_2}{dt} - k_2(x_2 - x_1) \qquad (1.4)$$

Assuming that we know d^2x_1/dt and d^2x_2/dt^2, we can integrate these variables twice to get x_1 and x_2, respectively. This approach has been used as a standard procedure with DC analog computers. Equations (1.4) allow us to construct the block diagram shown in Fig. 1.6.

In DC analog computers, since the number of stages in each operational amplifier was, and still is, an odd number (regularly 3 stages with a total gain of 100,000 or more), the sign at the output is reversed. In other words, a positive input v_1 to an add unit will give a negative output $-v_1$. The block diagram shown in Fig. 1.6 can be used as input to the simulation language CSMP- IBM 1130 version. On the other hand, if we want to use

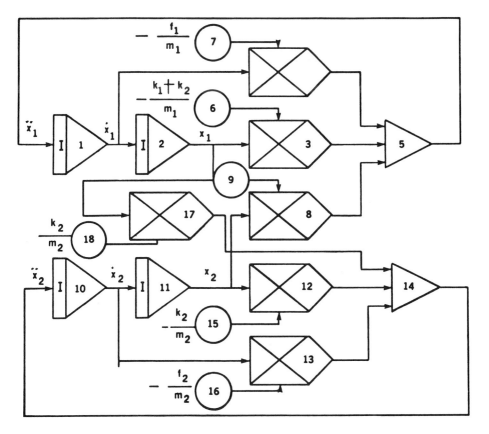

Figure 1.6   Block Diagram of Mechanical System.

a DC analog computer we need to make the necessary changes to consider the inversion of the sign at the output of each unit. The block diagram is really our model to input into the simulation language. It is clear by now that to use simulation languages such as the one mentioned, we need to go from our system to its block diagram representation. This block diagram constitutes the model of our system. This block diagram is used to implement the simulation.

3.2   Discrete Time Simulation

Discrete time simulation involves the simulation of a system that is described by a set of difference equations. We now present an example of this type of system.

Example 1.2   Fig. 1.7 shows a rotating commutator segment for connecting capacitor C_0 first to a sinusoidal voltage source and successively to capacitors C_1, C_2, and C_3. At $t=0$ the segment is on brush 4, and all four capacitors are without charge. After the commutator has made contact with brush 4, it goes to the other three brushes, namely, 1,2, and 3.

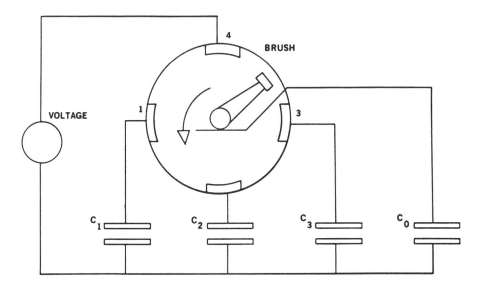

Figure 1.7   Rotating Commutator Segment.

To simplify our difference equations assume that

$$\frac{C_1}{C_0} = \frac{C_2}{C_0} = \frac{C_3}{C_0} = 0.1$$

$$\frac{\omega}{\beta} = 8$$

and

$$\text{Let } \alpha = \frac{1}{11}$$

Our problem is to write down the system of difference equations that define the voltages in the three capacitors C_1, C_2, and C_3.
The sampling interval is

$$T = \frac{T_1}{8} \tag{1.5}$$

Therefore, the voltage on capacitor C_0 for revolution n is given by

$$v_0(n) = V_m \sin(n\beta T) \tag{1.6}$$

For the first revolution we can write

$$v_1(1) = \frac{C_0}{C_1 + C_0} v_0(1) = 10 \, \alpha \, v_0(1) \tag{1.7}$$

Similarly for the other voltages

$$v_2(1) = 10 \, \alpha \, v_1(1) \tag{1.8}$$

$$v_3(1) = 10 \, \alpha \, v_2(1) \tag{1.9}$$

In general we can write

$$v_0(n) \, C_0 + v_1(n-1) \, C_1 = v_1(n) \, (C_0 + C_1) \tag{1.10}$$

Then, we can write the system of difference equations

$$\begin{cases} 10\alpha v_0(n) + \alpha v_1(n-1) = v_1(n) \\ 10\alpha v_1(n) + \alpha v_2(n-1) = v_2(n) \\ 10\alpha v_2(n) + \alpha v_3(n-1) = v_3(n) \end{cases} \tag{1.11}$$

This set of equations is now amenable for applying discrete time simulation techniques. Chapter 9 deals with such systems.

3.3   Discrete Event Simulation

This approach, as already discussed in some detail, deals with the mathematical modeling and simulation of systems subject to demands whose occurrence and lengths can, in general, be specified only probabilistically [1,2]. Examples of systems subject to these operational conditions are:

- *Telephone Systems* - The purpose of these systems is to provide the means of connecting two users of the system, when one of them requests the service. The facilities needed to establish and maintain a communications path between a pair of telephone sets are provided from a common pool, to be used by a call when required and returned to the pool when no longer needed. This introduces the probability that the system will be unable to set up a call on demand because of a lack of available equipment at that time. If we take into consideration the latter, the following question arises: *How much equipment must be provided so that the proportion of calls subject to delays will be below a specified acceptable level?*
- *Computer Systems* - A multiprogramming computer system with virtual memory creates an environment where transactions compete to receive service from the CPU. The CPU shares processing with all transactions in ready mode resident in the real memory. When a transactions is finished processing, another of the same type and class takes over. We have here again a similar situation, and we can ask the question: *How much equipment must be provided so that system response time will be below a specified acceptable level?*
- *Taxicabs in New York* - The taxicab flow in New York City can be thought of in a similar fashion. The question here is: *How many cabs are needed in New York City, so that the waiting time for getting a free cab is at an acceptable level?* Obviously, this problem has not been studied to date.

All these problems, although different in nature, share some common features, namely,

1.   Probabilistic description of processes
2.   Service time requirements
3.   Waiting time for service (queue)

Furthermore, the performance of each can be measured using the same performance indices, such as response time, queue length, throughput, util-

ization, etc. We now present an example of a system modeled as a
discrete event system.

   Example 1.3   Consider the problem of modeling a local area ring network consisting of a set of processors, each of which perform a certain
function. The purpose of the model is to investigate system performance.
The ring model for this network appears in Fig. 1.8. For this system, we
assume that messages (i.e., job requests) arrive over communications links
to a ring interface (RI) node where they are buffered until the token (trav-

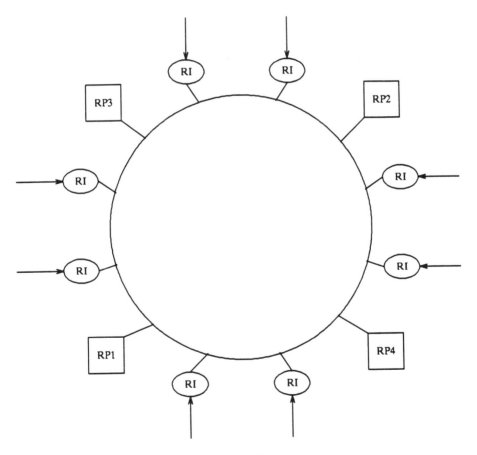

Figure 1.8   Local Area Ring Network Model.

eling around the ring) arrives. Once the token arrives, a specified number of messages enter the ring (ahead of the token) and travel the ring until reaching their destination processor on the ring (RP). For situations in which more than one RP can process a given class of messages, an incoming message (belonging to this class) is processed by one of these RPs based on some load sharing algorithm. Once the message has been processed, it awaits the arrival of the token, at which time it enters the ring ahead of the token and travels back to an appropriate RI node and from there back out of the ring. In this example there are four RPs and eight RIs. A central control processor is not taken into account for this study, and may or may not be present in the system. In any event we assume that there is a method for administering the system. The basic components of the model are the RPs, RIs, and network traffic that enters the ring at the RI nodes.

The model can be implemented using a general purpose discrete event simulation language. For this model, the component interactions carried out by the simulation program may be described as follows:

1. Traffic is assumed to follow a Poisson arrival process. The ring segment (defined as a portion of the ring between two RPs) on which an incoming message arrives is randomly selected from a user specified distribution function. Within a given ring segment, incoming messages are assigned a specific RI node using a uniform distribution. The destination RP is randomly selected from a user specified distribution function.

2. When a message to be processed at an RP node arrives at the ring, an entry point (RI node) to the ring is selected as described in Item 1. Once at the node, the message waits for the token to arrive and then begins its travel to the destination RP node.

3. If there is more than one RP node on the ring capable of processing the message, a load sharing algorithm is implemented.

4. When a message arrives at an RP, it is inspected to determine its destination. If the message is destined for the RP node, it is then processed by that node. If not, it is transmitted to the next RP on the ring. A message is accepted for processing if it belongs to the class of messages processed by the node and satisfies the load sharing algorithm requirements.

5. Once a message is accepted by the RP node, it joins a queue and awaits its turn for processing. The RP node is assumed to process one message (to completion) at a time. Once processing is completed, the

message is sent back to the ring and exits the ring at the point of entry.
6. Traffic flow on the ring is unidirectional.

The transmission time between two RI nodes can be calculated based on message size and ring transmission rate.

4. MODELING PROCESS

We now present a more complete view of the different steps that are involved in a modeling and simulation process. The modeling process, as presented by G. G. Hall [3], can be described as follows:

1. **Observe Situation** - The process or system to be modeled requires observation to fully understand not only its operation, but also its environment, and the impact that this environment has on our process or system. The learning process can be the most time-consuming aspect of the model development process. We need to determine the feasibility of developing the model. This study must be conducted with a clear idea about the purpose of our model (i.e., the experimental frame).

2. **Select Variables** - Once the process or system is well understood, we must select the variables that describe the process or system under study (abstraction). In carrying out this step, it may be convenient to use the *process or system decomposition* that will facilitate model development. This step is of great importance, since the selection of variables determines the level of detail with which the model will be described.

3. **MODEL - Relate Variables** - In this step we develop our model. The model decomposition mentioned in the previous step will facilitate this development stage. In performing this step, we are relating the variables already selected, as well as other parameters required specially for the particular model in question. If the previous steps are properly accomplished, this step is greatly facilitated.

4. **Validate** - Our model is completely developed, and now is the time when we need to check if it is a good representation of our process or system, as well as whether it will completely satisfy the model's purpose. This corresponds to the validation steps shown in Fig. 1.3. We first validate our model with the real system, we then verify that any code developed faithfully implements the model, and then compare the input/output behavior of our model with that of the system. If everything is satisfactory, we continue our modeling process. If not, we

need to go back to the beginning of our process and further observe our system. Obviously, this tells us that something was missed the first time.

5. **Predict** - The model is used to predict new results, which need to be evaluated within the process or system environment. The analysis of these results is the next step, new estimation.

6. **New Estimation** - The analysis of the prediction results allows us to obtain a new estimation of our process or system.

7. **Generalize** - The generalization of our model will emphasize our trust in it and at the same time may result in the development of new mathematical concepts.

8. **New Mathematics** - May be needed to accomplish model generalization. The impact of this generalization must be assessed through prediction.

9. **New Concept** - Sensitivity analysis of the model's behavior will allow us to state new concepts about our process or system.

5. SKILLS OF A MODELER

After reading through this chapter we might ask, *What are the skills required of a modeler?* G. G. Hall [3] lists the following attributes:

1. To Analyze Real Situations: actual data; abstract relevant parts.
2. To Model: invent relations; estimate consequences; asses results.
3. To Cooperate: work with experimentalists; interpret ideas; relate disciplines.
4. To Create: models; concepts; theories; mathematics.
5. To Communicate: confer; simplify and unify mathematics; sense of timing.
6. To Think Mathematically: mastery of mathematical ideas; broad training; initiate new structures.

Let us describe in more detail each of the above attributes. This will lead to a better understanding of the *skills* a modeler should posses, and how to reach that status.

To Analyze Real Situations - We must posses the required level of knowledge to understand the operation of the process to be modeled. To this end we need to analyze the *actual data*, and abstract from the process its relevant parts, which are directly related to the model's purpose.

To Model - invent relations; estimate consequences; asses results - To
model a complex process, we need to create relations among the variables involved. In doing this we need to define new variables/parameters
which are internal to the model itself. Sometimes we need to develop
new relations because there is no mathematical relation defined or just
because the mathematical calculation is too complex to be implemented
directly.

*To Cooperate - work with experimentalists; interpret ideas; relate discip-
lines* - In general the modeler needs to work with experimentalists and
learn from them about the operation, use, and behavior of the process or
system. In general, experimentalists are not trained for model development nor the use of mathematics for understanding the process or system. Therefore, it is up to the modeler to properly interpret the ideas
involved and the related disciplines.

To Create - models; concepts; theories; mathematics - The creation of
models involves a mixture of several concepts and theories, which must
be applied as appropriate. All these concepts and theories will use
mathematics as a link. It is important that the utilization of these concepts and theories be such that model generalization becomes a feasible
step.

To Communicate - confer; simplify and unify mathematics; sense of timing
- The model needs to be accepted by the community involved in the
analysis of the process or system. Therefore, model communication
must confer trust in it. The simplification and unity of mathematics are
important aspects of model development and need to be performed in an
effective way by the modeler. To be successful during model presentation it is important to have a good sense of timing.

*To Think Mathematically - mastery of mathematical ideas; broad training;
initiate new structures* - This statement is clear by itself. The modeler
should think about mathematics to the point of creating new structures.
The training in mathematics must be broad and deep, from algebraic
manipulations to multidimensional analysis of non-linear systems. If we
take into consideration this fact, the generalization of the model
developed will become more effective.

REFERENCES

[1] Ajmone Mardan M., Balbo G., and Conte G. - *Performance Models of Multiprocessor Systems*. Computer Systems Series, MIT Press, 1986.

[2] Buzen, J. P. - Queueing Network Models of Multiprogramming, Ph. D. Thesis, Division of Engineering and Physics, Harvard University, 1971.
[3] Hall, G. G. - Presentation at The First World Congress on Mathematics at the Service of Man. July 13-16, 1977, Barcelona, Spain.
[4] Korn G. A. and Korn T. M. - *Electronic Analog Computers*. McGraw-Hill Book Company, Inc.
[5] Saltzer J. H., Pogran K. T., and Clark D. D. - "Why a Ring?" *Computer Networks 7*, pp. 223-231, 1983.
[6] Zeigler B. P. - *Theory of Modelling and Simulation*. John Wiley & Sons, 1976.
[7] Zeigler B. P. - *Multifacetted Modelling and Discrete Event Simulation*. Academic Press, 1984.

2

Basic Concepts in Probability and Random Variables

Ricardo F. Garzia

AT&T Bell Laboratories
Columbus, Ohio

1. INTRODUCTION

This chapter will present basic concepts dealing with probability and random variables. These concepts form the basis for developing the necessary tools to deal with analytic and simulation models of **local computer networks**. There are several good books that address these topics; refer to [1-3].

Let us consider an event that exhibits only one of several possible outcomes $A_1, A_2, A_3, \ldots, A_{20}$. We can write this discrete event as:

$$A \rightarrow A_1, A_2, A_3, \ldots, A_{20} \tag{2.1}$$

There are no other outcomes for event A. Therefore, we can say that our system is exhaustive (complete). We also note that only one outcome can occur at a time, and that all of them are equally likely to occur.

We associate some of them with a success, and the rest with a failure, e.g., all the even events $A_2, A_4, \ldots, A_{20}$ are successes, while the odd events $A_1, A_3, \ldots, A_{19}$ are failures. Then,

$$\begin{cases} Success \rightarrow \{A_2, A_4, \ldots, A_{20}\} \\ Failure \rightarrow \{A_1, A_3, \ldots, A_{19}\} \end{cases} \tag{2.2}$$

Thus, success rate is given by

$$Success\ Rate\ =\ \frac{|\{A_2,A_4,....,A_{20}\}|}{|\{A_1,A_2,....,A_{20}\}|} \tag{2.3}$$

and failure rate is

$$Failure\ Rate\ =\ \frac{|\{A_1,A_3,....,A_{19}\}|}{|\{A_1,A_2,....,A_{20}\}|} \tag{2.4}$$

The success rate in our case is

$$Success\ Rate\ =\ \frac{10}{20}\ =\ 0.5$$

In other words, out of 20 different outcomes, 10 result in success and therefore 10 result in failure.

Concerning the above presentation, it is important to note:

1. The success rate is associated only with the possible outcomes of a given conceptual random experiment.
2. The conditions of the conceptual random experiment are given at the outset.
3. The exhaustive (complete) set of all possible outcomes can be obtained from the given conditions, and it is known.
4. The outcomes are mutually exclusive.
5. The outcomes are equally likely.
6. The event A is a compound event, which is reduced to a simple event only if one of the A_i is a success.

Now we can introduce the definition of probability: *Given an exhaustive and mutually exclusive set, on which all the outcomes can occur one at a time with equally likely occurrence, the probability of a success is the ratio of the number of successful outcomes to the total number of possible outcomes.* Thus,

$$Probability(Success)\ =\ \frac{number\ of\ successful\ outcomes}{total\ number\ of\ possible\ outcomes} \tag{2.5}$$

If we go back to the outcomes of event A, $A_1,A_2,A_3,....,A_{20}$, the probability of occurrence of any one of them is given by

$$P(A_i)\ =\ \frac{1}{20}\ =\ 0.05 \qquad (i=1,2,....,20) \tag{2.6}$$

Now, let us assume that we assign a different value to each outcome of the experiment, or

$$x_i = x(A_i) \tag{2.7}$$

we have thus defined a function x whose domain is the set $\{A_i\}$ of all outcomes. This function is called a *random variable*, and in this particular case it is a *discrete random variable* since it can only take on a countable number of values. For this discrete random variable, we have associated the probability of occurrence of each outcome (which in this case is always 0.05). In other words, we can write

$$\left\{ \begin{array}{ll} x_1(A_1) & P(A_1) \\ x_2(A_2) & P(A_2) \\ x_3(A_3) & P(A_3) \\ \quad \cdots\cdots \\ x_{20}(A_{20}) & P(A_{20}) \end{array} \right. \tag{2.8}$$

Since our system is mutually exclusive and exhaustive, we see that

$$\sum_{i=1}^{20} P(A_i) = 1.0 \tag{2.9}$$

We now introduce another probability function, the *distribution function*. To this end, we are interested in knowing what is the probability of obtaining the first i outcomes, namely, $A_1, A_2, \ldots, A_i$. This probability is given by

$$\sum_{j=1}^{i} P(A_j) = 1 - \sum_{j=i+1}^{20} P(A_j) \tag{2.10}$$

This probability will be designated with the capital letter F and defined as the *distribution function* or *probability distribution*; thus,

$$F(A_i) = \sum_{j=1}^{i} P(A_j) \tag{2.11}$$

The probability distribution is illustrated in Fig. 2.1.

We assume now that our random variable can take on any real value in the range $x_1 - x_2$. We define this type of random variable as a *continuous random variable*. The probability of this random variable is defined by the density function, which is illustrated in Fig. 2.2.

The probability that the continuous random variable takes on the value x is given by

$$P(x) = \rho(x)\,\Delta x \tag{2.12}$$

where $\rho(x)$ denotes the probability distribution of the continuous ran-

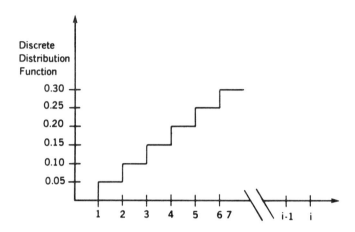

Figure 2.1   Discrete Distribution Function.

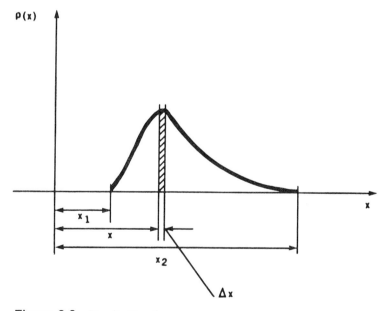

Figure 2.2   Density Function.

dom variable x, represented by the shaded area in the figure. It is obvious that since all the possible values of x are in the range

$$x_1 < x < x_2 \tag{2.13}$$

the following condition holds

$$\int_{x_1}^{x_2} \rho(\nu) \, d\nu = 1.0 \tag{2.14}$$

Similarly we can define the distribution function as

$$P(\xi < x) = F(x) = \int_{x_1}^{x} \rho(\nu) \, d\nu \tag{2.15}$$

We next review the three results that are closely related to the definition of probability.

1.1 Bernoulli's Theorem

Bernoulli's theorem states that *if in a series of M independent trials of a conceptual random experiment the number of successes of an event is N_M and the probability of the event is p, then the probability that the frequency ratio N_M/M differs from p by less than a for some positive ϵ, however small, tends to unity as M tends to infinity. In symbolic form,*

$$P \left[\left| \frac{N_M}{M} - p \right| < \epsilon \right] \rightarrow 1 \tag{2.16}$$

as $M \rightarrow \infty$.

1.2 Total Probability Postulate

A *random experiment* does not possess characteristics that can be treated mathematically. But, we know from experience that *random experiments* possess, in general, the characteristic of *statistical regularity*, i.e., the continuous tossing of a dice, the outcome of which portrays a random experiment, results in an average value after a countable set of experiments, thus showing an *statistical regularity*. Any *random experiment* which exhibits *statical regularity* will be called a *Conceptual Random Experiment*. With this definition we now proceed to the postulate of total probability which states that, *in a conceptual random experiment, if an event A can occur in the mutually exclusive forms A_1, A_2, A_3, ... , A_n, and in no other forms, then the probability of event A is the sum of the probabilities of events A_1, A_2, A_3, ... , A_n.* Symbolically,

$$P(A) = P(A_1) + P(A_2) + P(A_3) + \ldots + P(A_n) \tag{2.17}$$

What follows serves as motivation for a concrete case, not a general proof. To prove Equation (2.17), let k be the number of exhaustive, mutually exclusive, equally likely events in the experiment. Let $A_1, A_2, A_3, \ldots, A_n$ be the complete set of mutually exclusive forms of A; and let $m_1, m_2, m_3, \ldots, m_n$ be the respective number of simple events favorable to $A_1, A_2, A_3, \ldots, A_n$. The number of simple events favorable to A is

$$m_1 + m_2 + m_3 + \ldots + m_n \tag{2.18}$$

Consistent with the definition of probability, the probability of event A is

$$P(A) = \frac{m_1 + m_2 + m_3 + \ldots + m_n}{k} \tag{2.19}$$

Equation (2.17) can be written as follows

$$P(A) = \frac{m_1}{k} + \frac{m_2}{k} + \frac{m_3}{k} + \ldots + \frac{m_n}{k} \tag{2.20}$$

We recognize that

$$P(A_1) = \frac{m_1}{k}$$

$$P(A_2) = \frac{m_2}{k}$$

$$P(A_3) = \frac{m_3}{k}$$

$$\ldots\ldots\ldots$$

$$P(A_n) = \frac{m_n}{k}$$

Therefore, replacing the above into Equation (2.20), we obtain Equation (2.17).

1.3 Compound Probability Postulate

This postulate deals with the probability of the joint occurrence of two events in a random experiment. The essence of the postulate lies in the concept of *conditional probability*.

The postulate of compound probability states that *in a conceptual random experiment, the probability of the joint occurrence of events A and B*

(A ∩ B) is equal to the product of the unconditional probability of event A and the conditional probability (i.e., the probability of occurrence of an event, given the occurrence of a previous event) of event B given that event A has occurred, or the product of the unconditional probability of event B and the conditional probability of event A given that event B has occurred. Symbolically,

$$P(A \cap B) = P(A)\, P(B/A) = P(B)\, P(A/B) \qquad (2.21)$$

We discuss a simple problem to better understand conditional probability.

From a deck of cards (52 cards) we want to know the probability of obtaining 2 aces, without returning the first ace drawn from the deck. For the first draw, the probability of obtaining an ace is

$$p_1 = \frac{4}{52} = \frac{1}{13}$$

now, if one ace was drawn, the probability of obtaining the second ace is

$$p_2 = \frac{3}{51}$$

This probability p_2 is a conditional probability since it assumes that only three aces remain in the deck, which had a total of 51 cards.

Then, the probability of obtaining two aces is given by

$$p = p_1\, p_2 = \frac{1}{13} \frac{3}{51} = 0.0045$$

2. CONTINUOUS AND DISCRETE RANDOM VARIABLES

Continuous and discrete random variables play an important role in our study; we therefore discuss some important theorems. With this objective we are going to discuss the following theorems:

- Average of a Discrete Random Variable
- Average of Continuous Random Variables
- Moments of Random Variables
- Some General Forms of Average
- Variance of Random Variables
- Average of the Product of Random Variables
- Average and Average Square of the Sum of Random Variables
- Characteristic Function of Random Variables

2.1   Average of a Discrete Random Variable

Given a set of values a_1, a_2, a_3, ..., a_n, the average value is

$$m = \frac{a_1 + a_2 + a_3 + ... + a_n}{n} \tag{2.22}$$

The concept of an average value is applicable to random variables, and it is of primary importance in probability theory. Consider a random variable ξ with the probability distribution $P_\xi(x_i)$. Since we do not have a definite set of quantities for computing the average value when the variable is random, we cannot apply Equation (2.19). Nevertheless, a definition of the average of a random variable can be given in terms of the distribution. This is given by

$$\bar{\xi} = \sum_{i=-\infty}^{\infty} x_i \, P_\xi(x_i) \tag{2.23}$$

The above value is also known as the *mathematical expectation* of ξ.

2.2   Average of Continuous Random Variables

$$\bar{\xi} = \int_{-\infty}^{\infty} x \, P_\xi(x) \, dx \tag{2.24}$$

Assume that we define a function $f(x)$ which provides a function value for each value of the continuous random variable x. Under these conditions, we can define the mean value of f(x) as

$$\overline{f(\xi)} = \int_{-\infty}^{\infty} f(x) \, P_\xi(x) \, dx \tag{2.25}$$

2.3   Moments of Random Variables

Similarly, we can define the mean value of the moments of the random variable:

1. For a discrete random variable

$$\overline{\xi^m} = \sum_{i=-\infty}^{\infty} x_i^m P_\xi(x_i) \tag{2.26}$$

2. For a continuous random variable

$$\overline{\xi^m} = \int_{-\infty}^{\infty} x^m \, P_\xi(x) \, dx \tag{2.27}$$

2.4 Some General Forms of Average

We present some mean value expressions of a general form involving one random variable. Assume that the possible values of ξ are $x_1, x_2, x_3, \ldots, x_n$, $P_\xi(x_i)$ is the probability function of this discrete random variable, and let M be a sufficiently large number of trials. Then we can write

$$N_1 \quad \text{successes for} \quad f(\xi) = f(x_1)$$

$$N_2 \quad \text{successes for} \quad f(\xi) = f(x_2)$$

$$\ldots\ldots\ldots\ldots\ldots\ldots\ldots$$

$$N_i \quad \text{successes for} \quad f(\xi) = f(x_i)$$

$$\ldots\ldots\ldots\ldots\ldots\ldots\ldots$$

$$N_n \quad \text{successes for} \quad f(\xi) = f(x_n)$$

The empirical average value $f(\xi)$ is given by

$$f(\xi)|_{em.av.} = f(x_1)\frac{N_1}{M} + f(x_2)\frac{N_2}{M} + \ldots + f(x_n)\frac{N_n}{M} \qquad (2.28)$$

Replacing the frequency ratios by the probability distribution, we define the mean of $f(\xi)$ as

$$\overline{f(\xi)} = \sum_{i=1}^{n} f(x_i) \, P_\xi(x_i) \qquad (2.29)$$

And in general we can write

$$\overline{f(\xi)} = \sum_{-\infty}^{\infty} f(x_i) \, P_\xi(x_i) \qquad (2.30)$$

We can go further and consider the presence of a parameter in $f(\xi)$. Therefore, consider the function $f(t,\xi)$, in which t is a real and continuous parameter; the mean of the function with respect to the distribution of ξ is given by

$$\overline{f(t,\xi)}^{\xi} = \sum_{i=-\infty}^{\infty} f(t,x_i) \, P_\xi(x_i) \qquad (2.31)$$

For a continuous function we have

$$\overline{f(t,\xi)}^{\xi} = \int_{-\infty}^{\infty} f(t,x) \, P_\xi(x) \, dx \qquad (2.32)$$

2.5   Variance of Random Variables

An important characteristic associated with a random variable ξ is the variance σ_ξ^2 which is defined as

$$\sigma_\xi^2 = \overline{(\xi - \bar{\xi})^2} \tag{2.33}$$

The square root of the variance σ_ξ is called the *standard deviation* of ξ. Using Equation (2.31), we can write

$$\sigma_\xi^2 = \sum_{i=-\infty}^{\infty} (x_i - \bar{\xi})^2 P_\xi(x_i) \tag{2.34}$$

Since $\bar{\xi}$ is a constant, the expansion of the above equation gives

$$\sigma_\xi^2 = \sum_{i=-\infty}^{\infty} (x_i^2 - 2\bar{\xi}x_i + \bar{\xi}^2) P_\xi(x_i)$$

$$= \sum_{i=-\infty}^{\infty} x_i^2 P_\xi(x_i) - 2\bar{\xi} \sum_{i=-\infty}^{\infty} x_i P_\xi(x_i) + \bar{\xi}^2 \sum_{i=-\infty}^{\infty} P_\xi(x_i)$$

$$= \overline{\xi^2} - \bar{\xi}^2 \tag{2.35}$$

2.6   Average of the Product of Independent Random Variables

Two random variables ξ and ν are called statistically independent if the events $[\xi \leqslant x]$ and $[\nu \leqslant y]$ for any ξ and ν satisfy

$$P[\xi \leqslant x, \nu \leqslant y] = P[\xi \leqslant x] \, P[\nu \leqslant y]$$

Generalizing, we can say that the random variables $\xi_1, \xi_2, ..., \xi_m$ are statistically independent if the events

$$[\xi_1 \leqslant x_1], \ [\xi_2 \leqslant x_2] \ .... \ [\xi_m \leqslant x_m]$$

are pairwise independent for any pair $\xi_1, \xi_2, ..., \xi_m$. Now, assuming that we have two random variables ξ and η which are statistically independent, we can write

$$P_{\xi,\eta}(x,y;\tau) = P_\xi(x) \, P_{\eta/\xi}(x/y;\tau) \tag{2.36}$$

the conditional probability density becomes an unconditional density,

$$P_{\eta/\xi}(y/x;\tau) = P_\eta(y) \tag{2.37}$$

The above equation allows us to write

$$\overline{\xi\eta} = \int_{-\infty}^{\infty} \int_{-\infty}^{\infty} x \, y \, P_x(x) P_\eta(y) \, dx \, dy \tag{2.38}$$

The double integral may be expressed as

$$\overline{\xi\eta} = \int_{-\infty}^{\infty} y \, P_\eta(y) \, dy \int_{-\infty}^{\infty} x \, P_\xi(x) \, dx \tag{2.39}$$

Therefore,

$$\overline{\xi\eta} = \overline{\xi} \, \overline{\eta} \tag{2.40}$$

We have therefore shown that the mean of the product of two statistically independent random variables is the product of their mean values. Clearly this result can be extended to any finite number of mutually (pairwise) independent random variables.

2.7 Average and Average Square of the Sum of Random Variables

We can write

$$\overline{\xi+\eta} = \int_{-\infty}^{\infty} \int_{-\infty}^{\infty} (x+y) \, P_{\xi\eta}(x,y) \, dx \, dy \tag{2.41}$$

which may be expressed as

$$\overline{\xi+\eta} = \int_{-\infty}^{\infty} x \, [\int_{-\infty}^{\infty} P_{\xi\eta}(x,y) \, dy \,] \, dx$$
$$+ \int_{-\infty}^{\infty} y \, [\int_{-\infty}^{\infty} P_{\xi\eta}(x,y) \, dx \,] \, dy \tag{2.42}$$

But we know that

$$\int_{-\infty}^{\infty} P_{\xi\eta}(x,y) \, dx = P_\eta(y) \tag{2.43}$$

$$\int_{-\infty}^{\infty} P_{\xi\eta}(x,y) \, dy = P_\xi(x) \tag{2.44}$$

Replacing Equations (2.42) and (2.43) into Equation (2.41), and after some simplifications, we arrive at

$$\overline{\xi+\eta} = \overline{\xi} + \overline{\eta} \tag{2.45}$$

The mean of the sum of two random variables which may or may not be statistically independent is the sum of their means. This result can be generalized to any finite number of random variables.

$$\overline{\xi_1+\xi_2+\xi_3+....+\xi_n} = \overline{\xi_1} + \overline{\xi_2} + \overline{\xi_3} + ... + \overline{\xi_n} \tag{2.46}$$

Now, let us calculate the mean of the square of a sum of two random variables. We can write

$$\overline{(\xi+\eta)^2} = \int_{-\infty}^{\infty} \int_{-\infty}^{\infty} (x+y)^2 P_{\xi\eta}(x,y) \, dx \, dy \tag{2.47}$$

Equation (2.47) can be written as follows

$$\overline{(\xi+\eta)^2} = \int_{-\infty}^{\infty} x^2 dx \int_{-\infty}^{\infty} P_{\xi\eta}(x,y) \, dy + \int_{-\infty}^{\infty} y^2 dy \int_{-\infty}^{\infty} P_{\xi\eta} \, dx$$

$$+ 2\int_{-\infty}^{\infty} \int_{-\infty}^{\infty} xy \, P_{\xi\eta}(x,y) \, dx \, dy \tag{2.48}$$

Therefore,

$$\overline{(\xi+\eta)^2} = \overline{\xi^2} + \overline{\eta^2} + 2\bar{\xi}\,\bar{\eta} \tag{2.49}$$

2.8   Characteristic Function of Random Variables

Let us consider Equation (2.32), in which we define a special function, for example,

$$f(t,\xi) = e^{jt\xi} \tag{2.50}$$

Then we can write

$$\overline{e^{jt\xi}} = \int_{-\infty}^{\infty} e^{jtx} P_{\xi}(x) \, dx \tag{2.51}$$

In the theory of probability this average of a function of t is called the *characteristic function* of the probability density, written as

$$p_{\xi}(t) = \int_{-\infty}^{\infty} P_{\xi}(x) \, e^{jtx} dx \tag{2.52}$$

In the terminology of harmonic analysis the characteristic function, as in Equation (2.52), is the Fourier transform of the probability density function. It is a special transform because the function is real and non-negative and has an integral over $(-\infty,\infty)$ equal to one. Clearly, if $P_{\xi}(x)$ is even, the characteristic function is real. But in general $p_{\xi}(t)$ is complex. Now, we can write

$$P_{\xi}(x) = \frac{1}{2\pi} \int_{-\infty}^{\infty} p_{\xi}(t) \, e^{-jtx} dt \tag{2.53}$$

The most important use of the characteristic function is for calculating the moments.

2.8.1 Calculation of Moments

The expected value of a random variable ξ determines only the center of gravity of its density. A more complete application of the statistics of ξ is possible if we know its other *moments*, which are defined by

$$m = E[\xi^k] = \int_{-\infty}^{\infty} x^k f(x) \, dx$$

where $E[\xi^k]$ is the kth moment of the random variable ξ.
If we let $\bar{\xi} = E[\xi]$, then the *central moment* is defined by

$$\overline{\xi_c^k} = E[(\xi - \bar{\xi})^k] = \int_{-\infty}^{\infty} (x - \bar{\xi})^k f(x) \, dx$$

We can also define the *absolute moment* by

$$E[|\xi|^k] = \int_{-\infty}^{\infty} |x|^k f(x) \, dx$$

and the *generalized moment* by

$$E[(\xi - \alpha)^k] = \int_{-\infty}^{\infty} (x - \alpha)^k f(x) \, dx$$

where α is an arbitrary constant.
From Equation (2.52) we see that the nth derivative of $P_\xi(t)$ is given by

$$p_\xi^{(n)}(t) = j^n \int_{-\infty}^{\infty} x^n P_\xi(x) \, e^{jtx} dx \tag{2.54}$$

Letting t take the value 0 and rearranging, we get

$$\int_{-\infty}^{\infty} x^n P_\xi(x) \, dx = \frac{1}{j^n} p_\xi^{(n)}(0) \tag{2.55}$$

Then,

$$\overline{\xi^n} = \frac{1}{j^n} p_\xi^{(n)}(0) \tag{2.56}$$

From this result we can determine the moments of ξ through differentiation of the characteristic function.

2.9 Probability Density of a Sum of Statistically Dependent Variables

Let us assume that

$$\zeta = \xi + \eta \tag{2.57}$$

Then we can write

$$p_\zeta(t) = \overline{e^{j\zeta\zeta}} = \overline{e^{jt(\xi+\eta)^\zeta\eta}} \tag{2.58}$$

$$p_\zeta(t) = \int_{-\infty}^{\infty}\int_{-\infty}^{\infty} e^{jt(\xi+\eta)}\, p_{\xi\eta}(x,y)\, dx\, dy \tag{2.59}$$

If ξ and η are statistically independent, we replace the joint density by the product of the unconditional densities and obtain

$$p_\zeta(t) = \int_{-\infty}^{\infty}\int_{-\infty}^{\infty} P_\xi(x)\, P_\eta(y)\, e^{jt(\xi+\eta)}\, dx\, dy \tag{2.60}$$

$$= \int_{-\infty}^{\infty} P_\xi(x)\, e^{jt\xi}\, dx \int_{-\infty}^{\infty} P_\eta(y)\, e^{jt\eta}\, dy \tag{2.61}$$

Therefore,

$$p_\zeta(t) = p_\xi(t)\, p_\eta(t) \tag{2.62}$$

Generalizing, we have

$$\xi = \xi_1 + \xi_2 + \xi_3 + \dots + \xi_n \tag{2.63}$$

$$p_\xi(t) = p_{\xi_1}(t)\, p_{\xi_2}(t)\, p_{\xi_3}(t) \dots p_{\xi_n}(t) \tag{2.64}$$

Example 2.1 Consider the system shown in Fig. 2.3. The random variables η_1, η_2, and η_3 are applied as inputs, and are assumed to be independent. The density functions of these random variables are known and designated by $\eta_1(x)$, $\eta_2(x)$, and $\eta_3(x)$, Fig. 2.4.

The system selects, on a continuous basis, the random variable of maximum value. Let us denote by ρ the output of this process and by $\rho(x)$ its density function. The question now is to see if this is a deterministic or stochastic system.

We can express this process mathematically as follows:

$$\rho = \max(\eta_1,\eta_2,\eta_3)$$

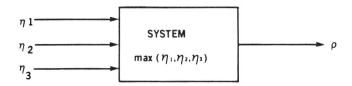

Figure 2.3 System.

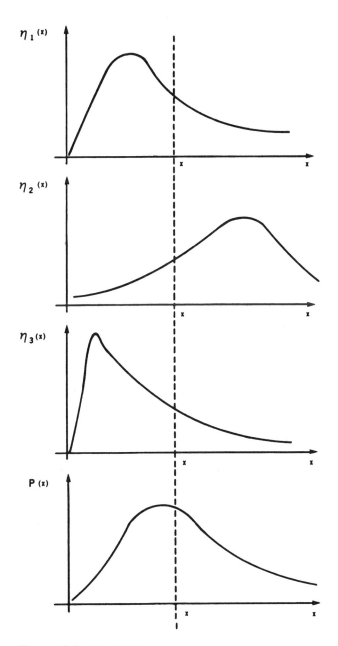

Figure 2.4 Distributions.

Suppose that $\rho = x$, for some real number x, then the probability that $\rho = \eta_1$ (i.e., $\eta_1 = x$ and $\eta_2 \leqslant x, \eta_3 \leqslant x$) is given by

$$Probability \ (\rho = \eta_1) \ = \ \eta_1(x) \ dx \int_{-\infty}^{x} \eta_2(v) \ dv \int_{-\infty}^{x} \eta_3(\phi) \ d\phi$$

But,

$$Probability \ (\rho = x) \ = \ \rho(x) \ dx$$

and so if $\rho = \eta_1$, we have

$$\rho(x) \ dx \ = \ \eta_1(x) \ dx \int_{-\infty}^{x} \eta_2(v) \ dv \int_{-\infty}^{x} \eta_3(\phi) \ d\phi$$

In general if we take into consideration the contribution of η_2 and η_3 to ρ, the probability that $\rho = x$ is given by

$$\begin{aligned}
\rho(x) \ dx \ = \ & \eta_1(x) \ dx \int_{-\infty}^{x} \eta_2(v) \ dv \int_{-\infty}^{x} \eta_3(\phi) \ d\phi \\
+ \ & \eta_2(x) \ dx \int_{-\infty}^{x} \eta_1(v) \ dv \int_{-\infty}^{x} \eta_3(\phi) \ d\phi \\
+ \ & \eta_3(x) \ dx \int_{-\infty}^{x} \eta_1(v) \ dv \int_{-\infty}^{x} \eta_2(\phi) \ d\phi
\end{aligned}$$

Therefore, we have

$$\rho(x) \ = \ \frac{d}{dx} [Q_1(x) Q_2(x) Q_3(x)]$$

where,

$$Q_j(x) \ = \ \int_{-\infty}^{x} \eta_j(v) \ dv$$

After integration we have

$$Q_\rho(x) \ = \ Q_1(x) \ Q_2(x) \ Q_3(x)$$

In general, for n random variables applied as input to the system, we can write

$$\rho \ = \ \max(\eta_1, \eta_2, \ldots, \eta_n)$$

$$Q_\rho(x) \ = \ \prod_{j=1}^{n} Q_j(x) \tag{2.65}$$

3. DISTRIBUTIONS OF RANDOM VARIABLES

Consider an experiment that can be described by a stochastic system. We assume that the realization of this experiment covers an infinite time, which will assure that all possible outcomes appear in the output of our stochastic system. Every time this experiment is performed, the outcome will be different, the difference is owing to the fact that the output samples appear in different sequences. If we perform this experiment an infinite number of times, we can have an ensemble made up of infinite output profiles, called members of the ensemble, as the one shown in Fig. 2.5.

Of course, in the previous explanation there is a strong idealization of the stochastic process: 1. we can not conduct our experiment in an interval of infinite duration; and 2. we cannot repeat our experiment an infinite number of times. But this idealized process leads us to understand an important characteristic of these stochastic systems which can be stated as follows: *The average value of the member profiles as a function of time is the same as the average value of the members across the ensemble, at any point in time.* This is an important characteristic of each particular ensemble. This type of stochastic process is called a stationary stochastic process or, in short, Ergodic process.

We now make a practical realization of this experiment, which will clarify the philosophy of the ensemble characteristics. Take any member of the ensemble and sample it in an infinite number of segments, as shown in Fig. 2.6.

We cut with a pair of scissors all these sample values and place them in a box. Now, every time that we want to perform our experiment, we start drawing from the box, one sample at a time, and place them down one after the other. When we have exhausted all the samples, we have a new member of our ensemble. Now, we are able to understand the philosophy behind our experiment: 1. Each member has all possible values (exhaustive system); 2. the order in which they are drawn is random.

If we consider the last two points, and compare them with the basic postulates of probability, we see that this ensemble possesses the basic characteristics stated above, and can be treated probabilistically.

We now present the probability distributions of two different types of random variables, namely,

• Discrete Random Variables
• Continuous Random Variables

Both of them play an important role in our network environment.

3.1 Distributions of Discrete Random Variables

With the framework presented in the preceding chapter on basic probability concepts and related theorems, the introduction of discrete and continuous

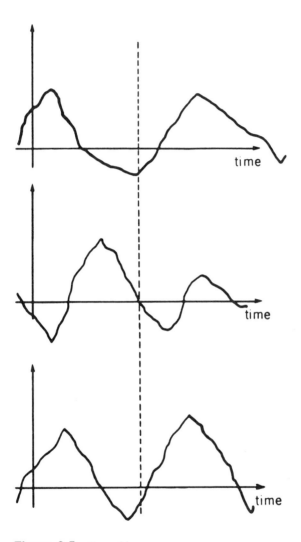

Figure 2.5   Ensemble.

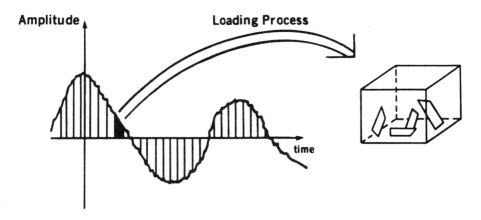

Figure 2.6 Sampling Process.

random variables, and the concept of Ergodic functions, we are ready now to study the distributions of the most common discrete random variables:

- Bernoulli Random Variables
- Binomial Random Variables
- Geometric Random Variables
- Poisson Random Variables
- Uniform Random Variables

For references on this topic see [1,4]. A discrete random variable is defined as the variable associated with a countable number of events. The probability is defined by

$$P[\xi_i]$$

where ξ_i is the discrete random variable.

3.1.1 Bernoulli Random Variables

The concept of Bernoulli random variables is of great importance because it allows a precise definition of the binomial and geometric random variables.

Several important discrete random variables are derived from the concept of a Bernoulli sequence of trials. A Bernoulli trial is a random experiment in which there are only two possible outcomes, usually *success* and *failure*, or *one* and *zero*, with the respective probabilities p and q. Of course we have that

$$p + q = 1 \tag{2.66}$$

A Bernoulli random variable ξ is completely determined by a parameter p, since from knowledge of this parameter and Equation (3.2) we can calculate the other

$$q = 1 - p \tag{2.67}$$

The expected value is p, and the $Var[\xi] = pq$.

3.1.2 Binomial Random Variables

Consider a Bernoulli sequence of n trials where the probability of success in each trial is constant and is given by p. A sequence of n Bernoulli trials can be represented by a string ξ_1, ξ_2, ξ_3, ..., ξ_n in which either ξ_i is a success or a failure. Assuming an integer k between 0 and n, we are interested in the probability of having k successes in n Bernoulli trials. The probability of having k successes and $n - k$ failures in n trials is

$$C_n^k = p^k q^{n-k} \tag{2.68}$$

The above equation is true since each trial is independent. The number of strings ξ_1, ξ_2, ξ_3, $\cdots$, ξ_n representing k successes and $n - k$ failures is just the number of ways the k indices representing success can be chosen from the n indices, that is C_k^n, which represent the number of combinations of n trials taken k at a time.

$$P(k,n,p) = \begin{bmatrix} n \\ k \end{bmatrix} p^k q^{n-k} \tag{2.69}$$

where $q = 1 - p$. Equation (2.69) can be written as follows

$$P(k,n,p) = \begin{bmatrix} n \\ k \end{bmatrix} p^k (1-p)^{n-k}$$

The expected value of ξ (mean) is given by

$$E[\xi] = E[\xi_1] + E[\xi_2] + E[\xi_3] + \dots + E[\xi_n] = np \tag{2.70}$$

since $E[\xi_i] = p$ for each i. The variance is given by

$$Var[\xi] = Var[\xi_1] + Var[\xi_2] + Var[\xi_3] + \dots$$
$$+ Var[\xi_n] = n\, p\, q = n\, p\, (1-p) \tag{2.71}$$

since $Var[\xi_i] = pq$ for each i, and the trials are independent.

3.1.3 Binomial Theorem

The binomial theorem deals with the expansion of the nth power of $a+b$; generally speaking, the term multinomial theorem considers the expansion of the nth power of the sum of the n numbers. For the binomial powers we have

$$(a+b)^n = \sum_{k=0}^{n} \begin{bmatrix} n \\ k \end{bmatrix} a^{n-k}b^k \tag{2.72}$$

where the coefficient

$$\begin{bmatrix} n \\ k \end{bmatrix} = C_k^n = \frac{n!}{k!\,(n-k)!} \tag{2.73}$$

The coefficient C_k^n has the following properties

$$\begin{bmatrix} n \\ k \end{bmatrix} = \begin{bmatrix} n \\ n-k \end{bmatrix} \tag{2.74}$$

$$\begin{bmatrix} n \\ 0 \end{bmatrix} = \begin{bmatrix} n \\ n \end{bmatrix} = 1 \tag{2.75}$$

$$\begin{bmatrix} n+1 \\ k \end{bmatrix} = \begin{bmatrix} n \\ k-1 \end{bmatrix} + \begin{bmatrix} n \\ k \end{bmatrix} \tag{2.76}$$

If $k \geqslant 2$ and n is a positive integer, then the multinomial theorem may be stated as

$$(a_1 + a_2 + a_3 + \ldots + a_k)^n = \sum \frac{n!}{k_1!k_2!k_3!\ldots.k_n!}\, a_1^{k_1}a^{k_2}a_3^{k_3}\ldots.a_k^{k_k} \tag{2.77}$$

where,

$$k_1 + k_2 + k_3 + \ldots + k_k = n \tag{2.78}$$

Example 2.2 The on-line computer system at AT&T Bell Laboratories, Columbus, Ohio, has 20 communication lines. The lines operate independently. The probability that any particular line is in use is 0.8. What is the probability that 10 or more lines would be in operation?

$$P[\xi \geqslant 10] = \sum_{k=10}^{20} \begin{bmatrix} 20 \\ k \end{bmatrix} (0.8)^k (0.2)^{20-k} = 0.99943$$

If the probability of each line being in use decreases to 0.5, the probability that 10 or more lines will be in operation is

$$P[\xi \geqslant 10] = \sum_{k=10}^{20} \begin{bmatrix} 20 \\ k \end{bmatrix} (0.5)^k (0.5)^{20-k} = 0.58809$$

Table 2.1 Busy Line Probability

Line prob.	$k=5$	$k=10$
.1	.04317	.000007
.2	.37035	.002596
.3	.76248	.047968
.4	.94904	.244667
.5	.99409	.588096
.6	.99968	.872477
.7	.99999	.982854
.8	.99999	.999437
.9	.99999	.999998

We may calculate the probabilities shown in Table 2.1.

3.1.4   Geometric Random Variables

Suppose that a sequence of Bernoulli trials is continued until the first success occurs. Let ξ be the random variable which counts the number of trials before the trial at which the first success occurs. Therefore, ξ can assume the values 0,1,2,3,... .ξ, it assumes the value 0 if and only if the first trial yields a success; hence, with probability p

$$P(k) = q^k p \tag{2.79}$$

The expected value is given by

$$E[\xi] = \frac{q}{p} \tag{2.80}$$

And the variance

$$Var[\xi] = \frac{q}{p^2} \tag{2.81}$$

3.1.5   Poisson Distributions

We are interested in presenting this probability distribution in some detail because it plays an important role in queuing network systems (this derivation was extracted from [3]). In Fig. 2.7 we show three members of a random experiment from which we are interested in finding the distribution of zero crossings during a time span τ. Therefore, we define the probability of having n zeros in time τ as

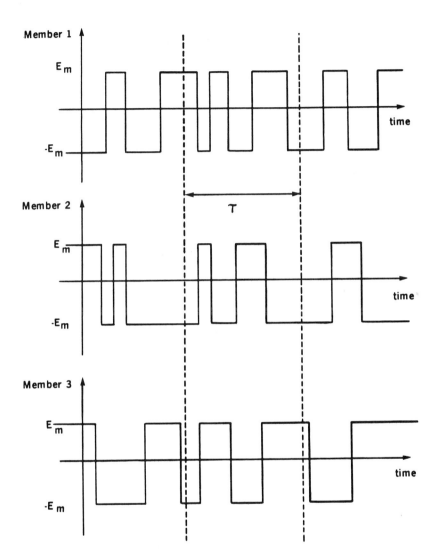

Figure 2.7 Members of the Ensemble.

$$P(\xi=n;\tau) = P_\xi(n;\tau) \qquad (2.82)$$

which will represent the probability distribution of interest. The expected value of the zeros in τ is given by

$$\bar{\xi} = \sum_{n=1}^{\infty} n\, P_\xi(n;\tau) = k\,\tau \qquad (2.83)$$

Thus, we define a new value for the expected value of the random variable k, which can be expressed as the expected value of the random variable per unit of time.

At this point we are interested in finding the expression for $P_\xi(n;\tau)$. First we need to find the probability of having a zero in $\Delta\tau$. To obtain this value and the associated one, not having a zero at $\Delta\tau$, let us divide the interval τ into m equal subintervals which we will designate as $\Delta\tau$. The length of these intervals is small enough so that only one zero can occur in each subinterval. This situation is shown in Fig. 2.8.

Therefore, we can define the outcomes of event A as follows

A_1 has a zero in τ, but one zero in $\Delta\tau$

A_2 has 2 zeros in τ, but one zero in $\Delta\tau$

A_3 has 3 zeros in τ, but one zero in $\Delta\tau$

$$\dots\dots\dots\dots\dots\dots\dots$$

A_m has m zeros in τ, but one zero in $\Delta\tau$

We can express the preceding conditions as:

$$P(A_1) = P(\xi=1;\tau,\eta=1;\Delta\tau)$$
$$P(A_2) = P(\xi=2;\tau,\eta=1;\Delta\tau)$$
$$P(A_3) = P(\xi=3;\tau,\eta=1;\Delta\tau)$$
$$\dots\dots\dots\dots\dots\dots\dots$$
$$P(A_m) = P(\xi=m;\tau,\eta=1;\Delta\tau)$$

Therefore, if we apply the total probability theorem, the probability of having a zero in $\Delta\tau$ is given by

$$P(A) = P(A_1) + P(A_2) + P(A_3) + \dots + P(A_m) \qquad (2.84)$$

The above equation can be written as

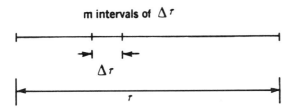

Figure 2.8 Span Time Partition.

$$P(A) = P(\xi=1;\tau,\eta=1;\Delta\tau) + P(\xi=2;\tau,\eta=1;\Delta\tau)$$
$$+ P(\xi=3;\tau,\eta=1;\Delta\tau)$$
$$+ \ldots + P(\xi=m;\tau,\eta=1;\Delta\tau)$$
$$(2.85)$$

and applying the compound probability theorem

$$P(\xi=1;\tau,\eta=1;\Delta\tau) = P(\xi=1;\tau) \, P(\eta=1;\Delta\tau)$$
$$P(\xi=2;\tau,\eta=1;\Delta\tau) = P(\xi=2;\tau) \, P(\eta=1;\Delta\tau)$$
$$P(\xi=3;\tau,\eta=1;\Delta\tau) = P(\xi=3;\tau) \, P(\eta=1;\Delta\tau)$$

$$\ldots\ldots\ldots\ldots\ldots\ldots\ldots\ldots$$

$$P(\xi=m;\tau,\eta=1;\Delta\tau) = P(\xi=m;\tau) \, P(\eta=1;\Delta\tau)$$

Equation (2.85) can now be written as follows

$$P(A) = [P(\xi=1;\tau) + P(\xi=2;\tau) + P(\xi=3;\tau)$$
$$+ \ldots + P(\xi=m;\tau)] \, P(\eta=1;\Delta\tau) \qquad (2.86)$$

Since we divide the time span τ into m equal parts, the probability of having a zero in any part is the same, and is given by

$$P(\eta=1;\Delta\tau) = \frac{\Delta\tau}{\tau} \qquad (2.87)$$

For us, the above expression is of major importance, since it expresses the probability of having a zero in a given interval, and does not depend on the location of this particular interval.

The probability of event A is given by

$$P(A) = \sum_{n=1}^{m} n\, P(\xi=n;\tau)\, \frac{\Delta\tau}{\tau} \qquad (2.88)$$

But, since we have already shown that $\sum n\, P_{\xi}(n;\tau)=k\tau$, we can write Equation (2.88) as follows

$$P(A) = k\tau\, \frac{\Delta\tau}{\tau} = k\, \Delta\tau \qquad (2.89)$$

which is a result we can obtain from the definition of k in Equation (2.83).

Likewise, the probability of having no zero in the interval $\Delta\tau$ is given by

$$P(B) = \sum_{n=1}^{m} P(\xi=n;\tau)\, (1-\frac{\Delta\tau}{\tau}) \qquad (2.90)$$

Now, for the next step in the Poisson probability distribution, we consider the time increment $\Delta\tau$ outside the τ interval, but touching the right-hand side, as shown in Fig. 2.9.

Now, the probability of having n zeros in the interval $\tau+\Delta\tau$ is given by

$$P(\xi+\eta=n;\tau+\Delta\tau) = P(\xi=n-1;\tau)\, P(\eta=1;\Delta\tau)$$
$$+\ P(\xi=n;\tau)\, P(\eta=0;\Delta\tau) \qquad (2.91)$$

but since

$$\begin{cases} P(\eta=1;\Delta\tau) = k\tau \\ P(\eta=0;\Delta\tau) = 1-k\tau \end{cases} \qquad (2.92)$$

Equation (2.91) becomes

$$P(\xi+\eta=n;\tau+\Delta\tau) = P(\xi=n-1;\tau)\, k\tau + P(\xi=n;\tau)\, (1-k\tau) \qquad (2.93)$$

The preceding equation can be written as follows

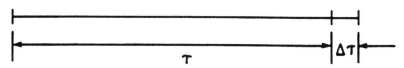

Figure 2.9 Location of $\Delta\tau$.

$$\frac{P(\xi+\eta=n;\tau+\Delta\tau) - P(\xi=n;\tau)}{\Delta\tau} - k\,P(\xi=n-1;\tau)$$

$$= -k\,P(\xi=n;\tau) \tag{2.94}$$

Taking the limit of Equation (3.31) as $\Delta\tau \to 0$, we recognize that the first term of the left member is the derivative of $P(n;\tau)$, or

$$\frac{d}{dt}P(n;\tau) = \lim_{\Delta\tau \to 0} \frac{P(\xi+\eta=n;\tau+\Delta\tau) - P(\xi=n;\tau)}{\Delta\tau} \tag{2.95}$$

The above equation has been written from the classic definition of the derivative formula

$$f'(x) = \lim_{\Delta x \to 0} \frac{f(x+\Delta x)-f(x)}{\Delta x} \tag{2.96}$$

We can write Equation (2.94) as follows

$$\frac{dP_\xi(n;\tau)}{d\tau} + k\,P_\xi(n;\tau) = k\,P_\xi(n-1;\tau) \tag{2.97}$$

The above equation is of the form

$$\frac{dy(t)}{dt} + R(x)y(t) - S(x) \tag{2.98}$$

with the solution given by

$$y(t) = e^{-\int R(x)dx} \int S(x)\, e^{\int R(x)dx} dx \tag{2.99}$$

Applying Equation (2.99) to the solution of Equation (2.97), we obtain

$$P_\xi(n;\tau) = e^{-k\tau} \int k\,P_\xi(n-1;\tau)\, e^{k\tau}\, d\tau \tag{2.100}$$

If $n=1$, Equation (2.100) becomes

$$P_\xi(1;\tau) = e^{-k\tau} \int k\,P_\xi(0;\tau)\, e^{k\tau}\, d\tau \tag{2.101}$$

We have to calculate the value of $P_\xi(0;\tau)$ to evaluate the integral of Equation (2.101). To this end, we need to obtain the probability of having no zeroes in the time span τ. Then, we can write

$$P(\xi+\eta=0;\tau+\Delta\tau) = P(\xi=0;\tau)\, P(\eta=0;\Delta\tau)$$

$$= P(\xi=0;\tau)\, (1-k\Delta\tau) \tag{2.102}$$

Equation (2.102) can be written as follows

$$\frac{P(\xi+\eta=0;\tau+\Delta\tau)-P(\xi=0)}{\Delta\tau} = -k\,P(\xi=0;\tau) \qquad (2.103)$$

Taking the limit of Equation (2.103) as $\Delta\tau \rightarrow 0$, we obtain the equation

$$\frac{dP_\xi(0,\tau)}{d\tau} = -k\,P_\xi(0;\tau) \qquad (2.104)$$

The solution of this equation is given by

$$P_\xi(0;\tau) = A\,e^{-k\tau} \qquad (2.105)$$

The value of the integration constant A is calculated from the condition $\tau=0$, or

$$P_\xi(0;0) = A = 1.0 \qquad (2.106)$$

Thus,

$$P_\xi(0;\tau) = e^{-k\tau} \qquad (2.107)$$

Therefore, with Equation (2.107), Equation (2.101) becomes

$$P_\xi(1;\tau) = e^{-k\tau}\!\int k\,e^{-k\tau}\,e^{k\tau}\,d\tau = k\tau\,e^{-k\tau} \qquad (2.108)$$

If $n=2$,

$$P_\xi(2;\tau) = e^{-k\tau}\!\int k^2\tau\,P_\xi(1;\tau)\,e^{k\tau}\,d\tau = \frac{(k\tau)^2}{2!}\,e^{-k\tau} \qquad (2.109)$$

and in general we can write

$$P_\xi(n;\tau) = \frac{(k\tau)^n}{n!}\,e^{-k\tau} \qquad (2.110)$$

which represents the probability distribution of Poisson. Fig. 2.10 shows the shape of these functions for different values of n.

The expected value of n for this distribution is given by

$$\bar{\xi} = \sum_{n=1}^{\infty} n\,P_\xi(n;\tau) = \sum_{n=1}^{\infty} n\,\frac{(k\tau)^2}{n!}\,e^{-k\tau} \qquad (2.111)$$

since the value of $e^{-k\tau}$ is not affected by the summation, Equation (2.111) can be written as

$$\bar{\xi} = e^{-k\tau} \sum_{n=1}^{\infty} n\,\frac{(k\tau)^n}{n!} \qquad (2.112)$$

or,

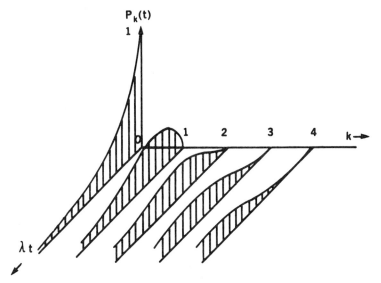

Figure 2.10   Poisson Distributions (copyright © 1975 Wiley reprinted by permission of John Wiley & Sons, Inc.).[1]

$$\bar{\xi} = e^{-k\tau} k\tau \left(1 + \frac{k\tau}{2!}2 + \frac{(k\tau)^2}{3!}3 + \frac{(k\tau)^3}{4!}4 + \cdots \right)$$

$$= e^{-k\tau} k\tau \left(1 + \frac{k\tau}{1!} + \frac{(k\tau)^2}{2!} + \frac{(k\tau)^3}{3!} + \cdots \right) \qquad (2.113)$$

Then,

$$E[\xi] = \bar{\xi} = e^{-k\tau} k\tau \, e^{k\tau} = k\tau \qquad (2.114)$$

as we expected and thus arrive at our definition, given by Equation (2.83).

The variance of this distribution is given by

$$Var[\xi] = \sigma_\xi^2 = \overline{(\xi - \bar{\xi})^2} \qquad (2.115)$$

which in words can be stated as *the mean of the squared difference of the values between the actual value of the discrete random variable and the mean value of the discrete random variable.*

Expanding the right-hand side of Equation (2.115), we have

---

[1] Reprinted from the book by L. Kleinrock - *Queueing Systems, Volume I: Theory*, Wiley, 1975.

$$\sigma_\xi^2 = \overline{\xi^2 - 2\xi\bar{\xi} + \bar{\xi}^2}$$

$$= \overline{\xi^2} - 2\bar{\xi}\bar{\xi} + \bar{\xi}^2 = \overline{\xi^2} - \bar{\xi}^2 \qquad (2.116)$$

Now, we know that $\bar{\xi} = k\tau$; therefore,

$$\bar{\xi}^2 = (k\ \tau)^2$$

but, we do not know $\overline{\xi^2}$. This is the second moment of the discrete random variable, given by

$$\overline{\xi^2} = \sum_{n=1}^{\infty} n^2\ P_\xi(n;\tau) \qquad (2.117)$$

Replacing $P_\xi(n;\tau)$, we obtain

$$\overline{\xi^2} = \sum_{n=1}^{\infty} n^2\ \frac{(k\tau)^n}{n!}\ e^{-k\tau} = e^{-k\tau} \sum_{n=1}^{\infty} n^2\ \frac{(k\tau)^n}{n!} \qquad (2.118)$$

Performing the expansion of the summation, we have

$$\overline{\xi^2} = e^{-k\tau}\ [k\tau + 2^2\ \frac{(k\tau)^2}{2!} + 3^2\ \frac{(k\tau)^3}{3!}$$

$$+\ 4^2\ \frac{(k\tau)^4}{4!} +\ \text{....}\ ]$$

$$= e^{-k\tau}\ k\tau\ [1 + 2\ \frac{k\tau}{1!} + 3\ \frac{(k\tau)^2}{2!}$$

$$+\ 4\ \frac{(k\tau)^3}{3!} +\ \text{....}\ ] \qquad (2.119)$$

Observing the series between brackets, we see that they may be written as

$$\frac{d}{d(k\tau)}[k\tau + \frac{(k\tau)^2}{1!} + \frac{(k\tau)^3}{2!} + \frac{(k\tau)^4}{3!} +\ \text{....}\ ]$$

$$= \frac{d}{d(k\tau)}\ k\tau\ [1 + \frac{k\tau}{1!} + \frac{(k\tau)^2}{2!}$$

$$+\ \frac{(k\tau)^3}{3!} +\ \text{....}\ ]$$

$$= \frac{d}{d(k\tau)}[k\tau\ e^{k\tau}] \qquad (2.120)$$

Replacing Equation (2.120) into Equation (2.119), we obtain

$$\overline{\xi^2} = e^{-k\tau} k\tau \frac{d}{d(k\tau)}[k\tau \, e^{k\tau}] \qquad (2.121)$$

and performing the derivation, we have

$$\overline{\xi^2} = e^{-k\tau} k\tau \, [e^{k\tau}+k\tau e^{k\tau}] = (k\tau)^2 + k\tau \qquad (2.122)$$

which expresses that the second moment of the Poisson discrete random variable is given by $(k\tau)^2 + k\tau$.

Now,

$$\sigma_\xi^2 = \overline{\xi^2} - \overline{\xi}^2 = k\tau \qquad (2.123)$$

Thus we see that the mean and variance of the Poisson distributions are identical and equal to $k\tau$.

3.1.6   Discrete Uniform Random Variables

A random variable ξ, which assumes a countable number of values $\xi_1, \xi_2, \xi_3, \cdots, \xi_n$, each with the same probability $1/n$ is called a discrete uniform random variable. The expected value is given by

$$E[\xi] = \frac{1}{n}\sum_{i=1}^{n} \xi_i \qquad (2.124)$$

The second moment is given by

$$E[\xi^2] = \frac{1}{n}\sum_{i=1}^{n} \xi^{2_i} \qquad (2.125)$$

and the variance can be calculated by

$$Var[\xi] = E[\xi^2] - (E[\xi])^2 \qquad (2.126)$$

3.1.7   Summary of Characteristics of Discrete Random Variables

Discrete random variables and their characteristics are summarized in the Table 2.2.

Example 2.3   The computer network configuration shown in Fig. 2.11 consists of N terminals connected to a Central Processor.

From experience we know that on the average, each terminal is used 70 percent of the time. Therefore, since for each line there is a Bernoulli distribution, we can write

$$p = 0.7$$

$$q = 0.3$$

Table 2.2 Discrete Random Variables

Random variable	Probability	$E(\xi)$	σ_ξ^2
Bernoulli	p	p	pq
Binomial	$P(k,n,p) = \begin{bmatrix} n \\ k \end{bmatrix} p^k q^{n-k}$	np	$np(1-p)$
Geometric	$P(k) = q^k p$	$\dfrac{q}{p}$	$\dfrac{q}{p^2}$
Poisson	$P_\xi(n,\tau) = \dfrac{(k\tau)^n}{n!}\, e^{-k\tau}$	$k\tau$	$k\tau$
Uniform	$P(\xi) = \dfrac{1}{n}$	$\dfrac{1}{n}\sum\limits_{i=1}^{n}\xi_i$	$E[\xi^2] - (E[\xi])^2$

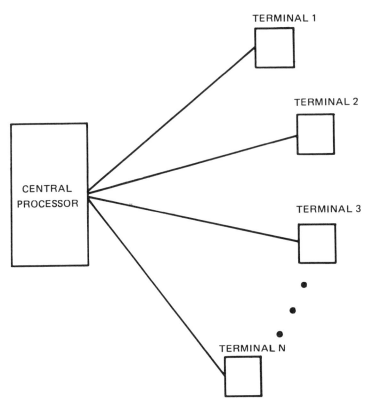

Figure 2.11   Central Processor Network.

The probability that no more than M terminals are in use, where $M \leqslant N$, is represented by a Binomial distribution

$$P = \sum_{k=1}^{M} \begin{bmatrix} n \\ k \end{bmatrix} p^k q^{N-k}$$

If $N = 25$, and $M = 15$, we have

$$P = \sum_{k=1}^{15} \begin{bmatrix} 15 \\ k \end{bmatrix} 0.7^k 0.3^{25-k} = 0.1894$$

And the probability that more than 15 terminals are in use is given by

$$P = \sum_{k=15}^{25} \begin{bmatrix} 25 \\ k \end{bmatrix} 0.7^k 0.3^{25-k} = 0.8106$$

The probability of the first 10 terminals being idle, and the 11th terminal being busy, is given by the geometric distribution as follows

$$P = 0.3^{10} \, 0.7 = 0.00000287$$

3.2 Distributions of Continuous Random Variables

A continuous random variable is characterized by the property

$$P[\xi = x] = 0 \tag{2.127}$$

for all real values of x within the range of the variable ξ.

For continuous random variables, we will be study the following distributions:

- Uniform random variables
- Exponential random variables
- Erlang-k random variables

3.2.1 Continuous Uniform Random Variables

A continuous uniform random variable has a distribution given by

$$f(t) = \frac{1}{b-a} \tag{2.128}$$

where a and b are as shown in Fig. 2.12.

The cumulative probability, also called distribution function, is given by

$$F(x) = \int_0^x f(v) \, dv = \frac{1}{b-a} \int_a^x dv = v \Big|_a^x = \frac{x-a}{b-a} \tag{2.129}$$

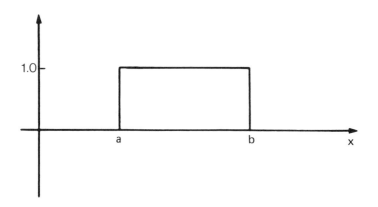

Figure 2.12   Uniform Random Distribution.

The expected mean value is given by

$$E[\xi] \;=\; \int_0^\infty \nu \, f(\nu) \, d\nu \;=\; \frac{1}{b-a} \int_a^b \nu \, d\nu$$

$$=\; \frac{1}{b-a} \left. \frac{x^2}{2} \right|_a^b \;=\; \frac{b+a}{2} \qquad\qquad (2.130)$$

The variance is given by

$$Var[\xi] \;=\; \sigma_\xi^2 \;=\; \int_0^\infty (\nu - E[\xi])^2 \, f(\nu) \, d\nu$$

$$=\; \frac{1}{b-a} \int_a^b \left[\nu^2 - 2\nu E[\xi] + \overline{E[\xi]}^2 \right] \;=\; \frac{(b-a)^2}{12}$$

$$(2.131)$$

Example 2.4   A disk drive in a computer system has a rotation time of 25 milliseconds. We would like to calculate the standard deviation of the position of the read/write head to read a particular record. We know that this distribution is of continuous uniform type, shown by its density in Fig. 2.13.

The variance of our random variable ξ is given by

$$Var[\xi] \;=\; \frac{25^2}{12} \;=\; 52.0833 \; milliseconds^2$$

Therefore, the standard deviation is given by

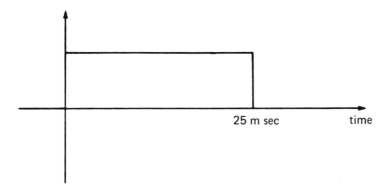

Figure 2.13   Time Distribution.

$$\sigma_\xi = (52.0833)^{0.5} = 7.2169 \; milliseconds$$

The probability that it will stop in any region on the track in 10 milliseconds is given by

$$P[\xi = 10] = \frac{10}{25} = 0.4$$

3.2.2   Exponential Random Variables

A continuous random variable ξ has an exponential distribution with parameter $\lambda > 0$ if its density function is given by

$$f(x) = \begin{cases} \lambda\, e^{-\lambda x} & x \geqslant 0 \\ 0 & x < 0 \end{cases} \tag{2.132}$$

The expected value of the exponential random variable is

$$E[\xi] = \int_0^x v\, f(v)\, dv = \int_0^x v\, \lambda\, e^{-\lambda v} dv = \frac{1}{\lambda} \tag{2.133}$$

In general, for the exponential random variable the expected value is designated by μ; thus,

$$\mu = \frac{1}{\lambda} \tag{2.134}$$

The probability function is given by

$$F(x) = \int_0^x \nu \ \lambda \ e^{-\lambda \nu} dv = 1 - e^{-\lambda x} \qquad (2.135)$$

The variance is

$$Var[\xi] = \sigma_\xi^2 = \int_0^\infty \left(\nu - \frac{1}{\lambda} \right)^2 \lambda \ e^{-\lambda \nu} dv = \frac{1}{\lambda^2} \qquad (2.136)$$

One reason for the importance of the exponential distribution in queuing theory is the *Markov property*, sometimes called *memoryless property*, which is stated as

$$P[\xi > t + h | \xi > t] = P[\xi > h] \qquad (2.137)$$

One interpretation of the above equation is that if ξ is the waiting time until a particular event occurs, and t units of time have produced no event, then the distribution of the waiting time left is the same as if no waiting time had passed; that is, the system does not remember that t time units have produced no arrival.

To prove the above property, we can write

$$P[\xi > t + h | \xi > t] \ P[\xi > t] = P[\xi > t + h] \ \bigcap \ P[\xi > t] \qquad (2.138)$$

which can be written as follows

$$P[\xi > t + h | \xi > t] = \frac{P[\xi > t + h] \ \bigcap \ P[\xi > t]}{P[\xi > t]} \qquad (2.139)$$

After some simplifications we obtain the proof of the memoryless property.

Example 2.5   Let us assume that engineering personnel use an on-line terminal to make routine calculations. If the time each engineer spends in a session at the terminal has an exponential distribution with an average value of 36 minutes, find

- The probability that an engineer spends less that 30 minutes at the terminal.
- More than one hour.
- If an engineer has already been at the terminal for 30 minutes, what is the probability that she or he will spend at least another hour at the terminal?.
- Ninety percent of the sessions end in less than R minutes. What is the value of R?

The mean service rate is given by

$$\lambda = \frac{1}{36}$$

Then the probability that an engineer spends less that 30 minutes is given by

$$\int_0^{30} \lambda\, e^{-\lambda x} dx = e^{-\lambda x}\Big|_{30}^{0} = 1 - e^{-\frac{30}{36}} = 0.565$$

more than one hour

$$\int_{60}^{\infty} \lambda\, e^{-\lambda x} dx = e^{-\lambda x}\Big|_{\infty}^{60} = e^{-\frac{60}{36}} = 0.189$$

For an engineer that already spent 30 minutes, the probability of spending another hour is

$$\int_{30}^{90} \lambda\, e^{-\lambda x} dx = e^{-\frac{30}{36}} - e^{-\frac{90}{36}} = 0.3427$$

The above calculation is incorrect. Can you tell why? The result should be 0.189. Can you explain?

The value of R can be obtained from

$$\int_0^{R} \lambda\, e^{-\lambda x} dx = 0.9 = e^{-\lambda x}\Big|_{R}^{0}$$

$$e^{-\frac{R}{36}} = 0.1$$

Therefore, the value of R is 82.8.

3.2.3 Erlang-k Distributions

This type of distribution is due to A. K. Erlang and has motivated the well known **Method of Stages**. Erlang recognized the simplicity of having the service times or interarrival times described by Poisson distributions, but he also recognized that this type of distribution could not be applied in all cases. Thus, Erlang conceived the notion of *decomposing* the service times or interarrival times into a collection of structured exponential distributions.

We consider the system indicated in Fig. 2.14(a). The service time is exponential and given by

$$\mu\, e^{-\mu t} \tag{2.140}$$

where μ is the service rate. The expected value and the standard deviation are given by

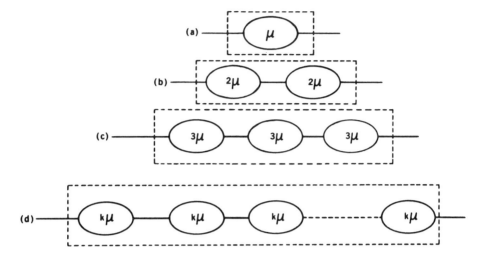

Figure 2.14   Erlang Distributions.

$$E[\xi] = \frac{1}{\mu} \qquad (2.141)$$

$$\sigma_\xi^2 = \frac{1}{\mu^2} \qquad (2.142)$$

Let us consider the system shown in Fig. 2.14(b). This system operates as follows: *The arriving transaction receives service from the first subsystem. When service is completed in this first subsystem, the transaction moves to the second subsystem and starts being serviced immediately. No transaction is accepted by the first subsystem until the second subsystem releases the transaction.* We see that this is a two stage service process. The transaction is processed in the first subsystem and then, when service is completed, the transaction moves to the second subsystem. No transaction can enter the system while any subsystem is servicing a transaction. To do the necessary calculations, we can cascade two subsystems with the same characteristics, as indicated in Fig. 2.15(a). The service time of each subsystem is given by

$$2\mu \, e^{-2\mu t} \qquad (2.143)$$

Therefore, the output of the system, using the real convolution theorem, can be described by

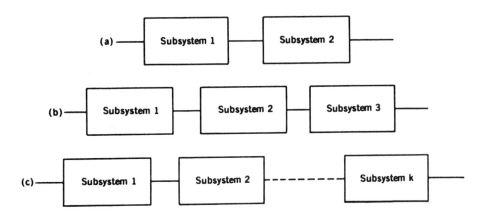

Figure 2.15 Subsystem Representation.

$$y(t) = \int_0^t 2\mu e^{-2\mu v} 2\mu e^{-2\mu(t-v)} \, dv$$
$$= 2 \mu \, (2\mu t) \, e^{-2\mu t} \tag{2.144}$$

The expected value and the variance are given by

$$E[\xi] = \frac{2}{\mu} \tag{2.145}$$

$$\sigma_\xi^2 = \frac{2}{\mu^2} \tag{2.146}$$

Consider the 3-stage system shown in Fig. 2.14(c). The service time provided by each subsystem can be described by

$$3\mu \, e^{-3\mu t}$$

Therefore, taking into consideration the system shown in Fig. 2.15(b), and Equation (2.144), we can write the output of subsystem 2 as

$$3\mu \, (3\mu t) \, e^{-3\mu t} \tag{2.147}$$

Then we can obtain the output of the third subsystem using the real convolution theorem, which is given by

$$y(t) = \int_0^\infty 3\mu \, (3\mu v) \, e^{-\mu v} 3\mu \, e^{-3\mu(t-v)} \, dv$$

$$= 3\mu \; \frac{(3\mu t)^2}{2} \; e^{-3\mu t} \tag{2.148}$$

and now generalizing, we can consider the k-stage system shown in Fig. 2.14(d). The service time provided by each subsystem can be described by

$$k\mu \; e^{-k\mu t} \tag{2.150}$$

Using the system represented in Fig. 2.15(c), we see that the output is given by

$$y(t) \; = \; \frac{k\mu \; (k\mu t)^{k-1}}{(k-1)!} \; e^{-k\mu t} \tag{2.151}$$

which is the Erlang-k distribution.

From a practical point of view, the physical model that Erlang had in mind was a service facility consisting of k independent stages, each with an exponential distribution of service time. He wanted this special facility to have the same average service time as a single facility whose service time is exponential with parameter μ.

$$F(t) \; = \; P[T \leqslant t]$$

$$= \; 1 \; - \; e^{-\mu k t} \; [1 \; + \; \frac{\mu k t}{1!} \; + \; \frac{(\mu k t)^2}{2!} \; + \; \cdots \; + \; \frac{(\mu k t)^{k-1}}{(k-1)!}] \tag{2.152}$$

Example 2.6 A given process in a computer network is performed in four stages. Each stage is known from experience to follow an exponential distribution with an average value of 12 minutes. We are interested in knowing:

1. What is the probability of completing processing during the first 30 minutes?
2. How does this probability vary in the range of 30 to 110?

On the average, since we have four stages, $\mu = 12$, applying Equation (2.152), we have

$$F(t) \; = \; P[T \leqslant t] \; = \; 1 \; - \; e^{-\frac{t}{12}} \; [1 \; + \; \frac{t}{12} \; + \; \frac{t^2}{2(12)^2} \; + \; \frac{t^3}{6(12)^3}]$$

Then, for $t = 30$ minutes, we have

$$F(30) \; = \; 1 \; - e^{-\frac{30}{12}} \; [1 \; + \; \frac{30}{12} \; + \; \frac{30^2}{2 \cdot 12^2} \; + \; \frac{30^3}{6 \cdot 12^3}] \; = \; 0.2426$$

Table 2.3 Processing
Time Probabilities

t	Probability
30	.2426
40	.4270
50	.5983
60	.7349
70	.8332
80	.8991
90	.9408
100	.9662
110	.9811

Calculating in the same way, we obtain the results shown in Table 2.3.

REFERENCES

[1] Allen, A. O. - *Probability, Statistics, and Queueing Theory with Computer Applications*, Academic Press, 1978.

[2] Cooper, R. B. - *Introduction to Queueing Theory, Second Edition*, Elsevier North Holland, Inc., 1981.

[3] Lee, Y. - *Statistical Theory of Communications*, Wiley, Inc., 1960.

[4] Melamed, B. - On Poisson Traffic Processes in Discrete State Markovian Systems with Applications to Queuing Theory, Technical Report 77-7, Department of Industrial and Operations Engineering, University of Michigan, 1977.

3

Compartmental Models

Mario R. Garzia

AT&T Bell Laboratories
Columbus, Ohio

In this chapter we present a brief look at a class of models that have found extensive use in the biological and chemical sciences [7,8,19]. A special case of these, known as Markov models, have also found wide use in hardware reliability [20] and queueing theory [17]. The reason for discussing compartmental models here is that we shall make use of them in Chapter 9 for modeling nonhierarchical communications networks. This application will make use of both linear and nonlinear compartmental models. The application of compartmental analysis to the study of communications networks dynamics is not common. Aside from special cases such as the Markovian models mentioned, compartmental analysis is seldom used in engineering. Quoting from [7], *"compartmental system analysis has a largely unexploited potential in non-biomedical engineering."* As will be seen in the following sections, several results are available concerning such models that can be directly applied to the study and understanding of network dynamics. Proofs to the results on compartmental analysis are not provided since this is beyond the scope of this introductory discussion. These proofs may be found in the sources cited.

Compartmental systems are a special class of dynamical systems. In general terms, dynamical systems are those systems which evolve as a function of time. The simplest continuous time dynamical system can be represented by the differential equation

$$\dot{x}(t) = ax(t) \tag{3.1}$$

with t, $x(t)$, $a \in \mathbb{R}$. The solution to this equation is given by

$$x(t) = Ce^{at} \tag{3.2}$$

where C is a constant determined by the initial condition (initial state) of the system. We can generalize this system by letting x be a vector. In this case we have a system of differential equations which can be written as

$$\dot{x} = Ax(t) \tag{3.3}$$

where $x \in \mathbb{R}^n$ and $A \in \mathbb{R}^{n \times n}$. We may also add a forcing term to the homogeneous system (3.3), resulting in the system

$$\dot{x} = Ax(t) + Bu(t) \tag{3.4}$$

with $u(t) \in \mathbb{R}^p$ and $B \in \mathbb{R}^{n \times p}$. $Bu(t)$ is the forcing term of the non-homogeneous system (3.4). More precisely, as in [9], we define a dynamical system as: *A dynamical system Σ is a mathematical object defined by the following axioms:*

i. *There is a time set T, a state set X, a set of input values U, a set of input functions $\Omega = \{u:T \rightarrow U\}$, a set of output values Y, and a set of output functions $\Gamma = \{\gamma : T \rightarrow Y\}$.*

ii. *T is an ordered subset of the real numbers.*

iii. *Ω satisfies the conditions*

   a. *Ω is nonempty.*

   b. *An input segment $u(t)$, $t_1 < t \leqslant t_2$ in Ω is restricted to $(t_1, t_2) \cap T$. If $u, u' \in \Omega$ and $t_1 < t_2 < t_3$, there exists $u'' \in \Omega$ such that $u''(t) = u(t)$, $t_1 < t \leqslant t_2$ and $u''(t) = u'(t)$, $t_2 < t \leqslant t_3$.*

iv. *There is a state transition function $\phi:T \times T \times X \times \Omega \rightarrow X$ whose value is the state $x(t) = \phi(t;\tau,x,u)$ in X, resulting at time $t \in T$ from the initial state $x(\tau) \in X$ at initial time $\tau \in T$, under action of the input $u \in \Omega$. ϕ has the following properties:*

   a. *ϕ is well defined for all $t \geqslant \tau$ but not necessarily for all $t < \tau$.*

   b. *$\phi(t;\tau,x,u) = x$ for all $t \in T$, $x \in X$, $u \in \Omega$.*

   c. *For all $t_1 < t_2 < t_3$ we have $\phi(t_3;t_1,x,u) = \phi(t_3;t_2,\phi(t_2;t_1,x,u),u)$ for all $x \in X$, $u \in \Omega$.*

   d. *If $u,u' \in \Omega$ and $u(t) = u'(t)$, $t < t \leqslant \bar{t}$, then $\phi(\bar{t};\tau,x,u) = \phi(\bar{t};\tau,x,u')$.*

v. *There is a readout map* $\eta:T \times X \to Y$ *that assigns the output*
    $y(t) = \eta[t,x(t)]$ *at time t.*

The study of such dynamical systems falls within the area of mathematical systems theory for which an extensive literature exists; see for instance [6,9,14,15] and the references therein. For the dynamical system (3.4), the state transition matrix ϕ is given by

$$\phi(t;0,x,u) = e^{At}x_0 + \int_0^t e^{A(t-s)}Bu(s)ds \qquad (3.5)$$

A dynamical system Σ is called linear if $\phi(t;\tau,.,.)$ is linear. That is, if for $\gamma_1,\gamma_2,\alpha_1,\alpha_2 \in \mathbb{R}$, we have

$$\begin{aligned}
\phi(t;\tau,\alpha_1 x_1 + \alpha_2 x_2, \gamma_1 u_1 + \gamma_2 u_2) &= \alpha_1\gamma_1\phi(t;\tau,x_1,u_1) \\
&+ \alpha_1\gamma_2\phi(t;\tau,x_1,u_2) \\
&+ \alpha_2\gamma_1\phi(t;\tau,x_2,u_1) \\
&+ \alpha_2\gamma_2\phi(t;\tau,x_2,u_2)
\end{aligned}$$

It can be shown, based on the definition of dynamical system, that a dynamical system is linear if and only if it can be represented by the system of equations (3.4) [15].

Dynamical systems can possess many properties of interest. For example, a system is said to be completely controllable if one can choose a suitable control function $u(t)$ which drives the system to the origin in finite time, given some initial state $x(0)$. A system is said to be completely observable if the value of the state vector can be determined from observations of the system output. This system feature is of interest when the state of the system can only be measured indirectly through the output function (or readout map), say,

$$y(t) = Cx(t) \qquad (3.6)$$

Another feature of interest concerns stability. In general, for a dynamical system of the form

$$\dot{x}(t) = f[x(t)] \qquad (3.7)$$

we define an equilibrium point $\hat{x}$ as any point in the domain of f such that $f(\hat{x})=0$. An equilibrium point $\hat{x}$ is called a stable equilibrium point if for each neighborhood N of $\hat{x}$ there exists a neighborhood $N' \subset N$ such that every solution $x(t)$ of (3.7), with initial condition in N', lies within N for all $t>0$. A stable equilibrium point is called asymptotically stable if it is an attractor. That is, $\lim_{t \to \infty} x(t) = \hat{x}$. Thus, our simple dynamical system (3.1) is asymptotically stable, with $\hat{x} = 0$, whenever $a < 0$. More generally, the system described by equation (3.3) is asymptotically stable if all eigenvalues of A have a negative real part.

The notions of stability discussed so far have dealt with homogeneous systems. For non-homogeneous systems such as the one in equation (3.4) another notion of stability becomes important, namely, bounded-input-bounded-output (b.i.b.o.) stability. A system is said to be b.i.b.o. stable if any bounded input results in a bounded output [4]. The following two results, appearing in [4], make a connection between the two notions of system stability discussed.

Theorem 3.1 [4]   - If system (3.3) is asymptotically stable, then the corresponding system described by equations (3.4) and (3.6) is b.i.b.o. stable.

Theorem 3.2 [4]   - If the system described by equations (3.4) and (3.6) is completely controllable, completely observable, and b.i.b.o. stable, then the corresponding system (3.3) is asymptotically stable.

With this brief review of dynamical systems as background, we now turn to the special subclass of compartmental systems, which will serve as background for our work in Chapter 9 dealing with the modeling of network dynamics.

1. COMPARTMENTAL ANALYSIS BACKGROUND

A standard description of material transfer processes is based on the concept of compartments and flows between compartments. If a "substance" is present in a system in several distinguishable forms or locations, then all the substance in a particular form and location can be said to constitute a compartment. Many compartments can coexist in the same location in a given system as, for instance, in the case of several different chemicals in a well-stirred beaker. The treatment of a compartment as a single store is an idealization, since a compartment may be a complex entity with more detailed internal structure. Furthermore, the concentration of a substance in the system is generally not uniform. A detailed distributed parameter model can be developed that takes into account the concentration of substances at each point in space. This approach leads to a model consisting of a set of partial differential equations that is, in general, a much harder problem to solve. In cases where the assumption of evenly distributed sub-

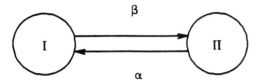

Figure 3.1 Two Compartment System.

stances is acceptable, compartmental models can be employed. This assumption, however, is not always appropriate and care must be taken in practice to assure that it is a reasonable one [1]. A simple two compartment model appears in Fig. 3.1. For this example, the substance in compartment I flows into compartment II at a rate β, and the substance in compartment II flows into compartment I at a rate α. This figure represents a closed system, that is, one in which substances are not exchanged with the environment.

Consider as an example the reversible chemical reaction

$$A + B \overset{\rightarrow}{\underset{\leftarrow}{}} AB$$

between two substances A and B, and let compartment I denote the concentration of AB, denoted $[AB]$, and compartment II denote the concentration of disassociated A, denoted $[A]$. Assume that the concentration of substance B is large enough so as not to be a factor in the reaction rate. Further assume that the transformation of A and B into AB is proportional to $[A]$, and that the transformation from $[AB]$ into A and B is proportional to $[AB]$. Then we have the following set of differential equations describing our system

$$\frac{d[AB]}{dt} = \alpha[A] - \beta[AB] \tag{3.8}$$

$$\frac{d[A]}{dt} = \beta[AB] - \alpha[A]$$

where α and β are the constants of proportionality. Models of this type have found extensive use in the study of chemical processes and in the biosciences [8].

2. LINEAR COMPARTMENTAL MODELS

A general linear compartmental model, with n compartments, is represented by the set of ordinary differential equations

$$\frac{dx_i}{dt} = b_i + \sum_{\substack{j=1 \\ j \neq i}}^{n} a_{ij}x_j - \sum_{\substack{j=1 \\ j \neq i}}^{n} a_{ji}x_i - a_{0i}x_i \qquad (3.9)$$

where $t \geqslant 0$, $x_i(0) = x_{0i}$, and $i = 1,2,\dots,n$. In this formulation, x_i is the amount of "material" in compartment i, and the a_{ij}'s are constant. The x_i's therefore describe the state of the system. Furthermore, a_{ij} is the flow rate to compartment i from compartment j, a_{ji} represents the flow rate from compartment i to compartment j, for $i \neq j$ and $i,j = 1,2,\dots,n$. The rate a_{0i} denotes the flow rate from compartment i to the environment. The flow rate from the environment into compartment i is given by b_i. It is clear from our discussion that compartmental models must satisfy $a_{ij} \geqslant 0$ and $b_i \geqslant 0$, for $i,j = 1,2,\dots,n$. For a system to be realizable as a linear compartmental model it is necessary and sufficient that $x_i(0) \geqslant 0$ and $a_{ij} \geqslant 0$ for $i = 0,1,\dots,n$; $j = 1,2,\dots,n$; $i \neq j$, and $a_{ii} \leqslant 0$. Fig. 3.2 presents the exchange between two compartments of this general, linear, n-compartment model. It is easy to show that linear, time-invariant compartmental systems, with constant $B = (b_i) \geqslant 0$, satisfy $x_i(t) \geqslant 0$ for all t and i whenever $x_i(0) \geqslant 0$ for all i. For the solution to be bounded it is necessary that $a_{ii} \leqslant 0$ [13]. These results also hold for time-varying $a_{ij}(t)$ and $B(t) \geqslant 0$ [13].

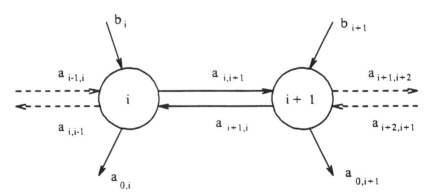

Figure 3.2 General Linear Compartmental Model.

We can rewrite equation (3.9) as

$$\dot{x}(t) = Ax(t) + Bu(t) \tag{3.10a}$$

$$y = Cx(t) \tag{3.10b}$$

where $x(t) = [x_1(t), x_2(t), ..., x_n(t)]^T$ denotes the state vector, $u(t)$ denotes the input (or control) function

$$a_{ii} = - (\sum_{\substack{j=1 \\ i \neq j}} a_{ij} + a_{oi})$$

and $A = (a_{ij})$ and $B = (b_i)$. We have also added the output map $y(t)$ through which we can observe the state $x(t)$ of the system. This is the standard form for a linear dynamical system with constant coefficients as discussed earlier.

Before starting our discussion on the properties of linear compartmental systems, we present two examples that are widely used in engineering and queueing theory. These examples belong to a special class of compartmental models known as Markovian models. As with compartmental models in general, Markovian models can have either continuous or discrete states, and evolve over continuous or discrete time. This gives rise to four types of Markov models, one for each combination of state and time. The discrete state continuous time Markov models that we shall be dealing with are known as Markov Processes. For all these models we assume that "objects" move independently from each other, and that transitions from one compartment to another are Markovian, that is,

$$Pr\{a < z_t \leqslant b \,|\, z_{t_1} = z_1, ..., z_{t_n} = z_n\} = Pr\{a < z_t \leqslant b \,|\, z_{t_n} = z_n\}$$

whenever $t_1 < t_2 < \cdots < t_n < t$, where $Pr\{z = b\}$ denotes the probability that the random variable z takes on the value b.

Example 3.1 **Markov Process** - Markov processes form a useful class of linear compartmental models which are widely used in reliability theory [20] and communications theory [17]. Fig. 3.3 presents a three state Markov model used to study the hardware availability/reliability of two identical systems operating in parallel, only one of which needs to be operational at any point in time for the entire system to be operational. An example of such a system is a duplex computer system (two identical computers working in parallel) such as those used in most central switching offices. In this case only one computer is needed to run the central office; the other operates as a backup if the first one malfunctions.

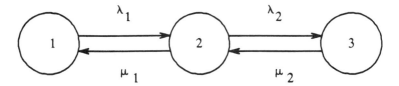

Figure 3.3 Three-State Markov Process.

Considering the interactions between the various compartments of Fig. 3.3, we obtain the associated system of differential equations

$$\dot{p}_0(t) = -\lambda_1 p_0(t) + \mu_1 p_1(t)$$

$$\dot{p}_1(t) = \lambda_1 p_0(t) - (\mu_1 + \lambda_2)p_1(t) + \mu_2 p_2(t)$$

$$\dot{p}_2(t) = \lambda_2 p_1(t) - \mu_2 p_2(t)$$

with initial conditions

$$p_0(0) = 1, \ p_1(0) = p_2(0) = 0$$

For this model, state $p_0(t)$ represents the probability that both systems are in operation. State $p_1(t)$ represents the probability that one of the two systems is out of service, and $p_2(t)$ represents the probability that both systems are out of service. Clearly, the system is not operational when it is in state $p_2(t)$. The parameters λ_1, λ_2 and μ_1, μ_2 represent the failure and repair rates, respectively. The above set of differential equations describes the probabilistic transitions between states. In developing this model we assume that the transition probabilities are constant, the probability of transition in time Δt from one state to another is $\lambda_i \Delta t$ (or $\mu_i \Delta t$) for appropriate i, and that the probability of more than one transition in time Δt can be neglected. The initial conditions selected reflect the assumption that we start with a fully operational system. If we let

$$P(t) \triangleq \begin{bmatrix} p_0(t) \\ p_1(t) \\ p_2(t) \end{bmatrix}$$

and

$$A \triangleq \begin{bmatrix} -\lambda_1 & \mu_1 & 0 \\ \lambda_1 & -(\lambda_2+\mu_1) & \mu_2 \\ 0 & \lambda_2 & -\mu_2 \end{bmatrix}$$

then we can write the system of differential equations equivalently as

$$\dot{P}(t) = A\,P(t)$$

Taking the Laplace transform of $P(t)$ on both sides, denoted by $\hat{P}(s)$, and rearranging terms, we obtain the following linear system of equations

$$s\hat{P}(s) - A\hat{P}(s) = P(0)$$

which can be solved to yield

$$\hat{P}(s) = \frac{1}{\Lambda} \begin{bmatrix} s^2+(\mu_1+\mu_2+\lambda_2)s+\mu_1\mu_2 \\ \lambda_1 s+\lambda_1\mu_2 \\ \lambda_1\lambda_2 \end{bmatrix}$$

where

$$\Lambda = s[s^2+(\lambda_1+\lambda_2+\mu_1+\mu_2)s+(\lambda_1\mu_2+\mu_1\mu_2+\lambda_1\lambda_2)]$$

In steady state, using the initial value theorem for Laplace transforms, we have

$$P \triangleq \lim_{t\to\infty} P(t) = \lim_{s\to 0} s\hat{P}(s)$$

$$= \frac{1}{\lambda_1\mu_2+\mu_1\mu_2+\lambda_1\lambda_2} \begin{bmatrix} \mu_1\mu_2 \\ \lambda_1\mu_2 \\ \lambda_1\lambda_2 \end{bmatrix}$$

To obtain the solution to our system of linear differential equations we can take the inverse Laplace transform of $\hat{P}(s)$ to arrive at

$$P(t) = \begin{bmatrix} \mu_1\mu_2/s_1 s_2 \\ \lambda_1\mu_2/s_1 s_2 \\ \lambda_1\lambda_2/s_1 s_2 \end{bmatrix}$$

$$+ \begin{bmatrix} (s_1{}^2+(\mu_1+\mu_2+\lambda_2)s_1+\mu_1\mu_2)/s_1(s_1-s_2) \\ \lambda_1(s_1+\mu_2)/[s_1(s_1-s_2)] \\ \lambda_1\lambda_2/[s_1(s_1-s_2)] \end{bmatrix} e^{s_1 t}$$

$$
+ \left[\begin{array}{c} (s_2{}^2 + (\mu_1 + \mu_2 + \lambda_2)s_2 + \mu_1\mu_2)/s_2\,(s_2 - s_1) \\[1mm] \lambda_1(s_2 + \mu_2)/[s_2\,(s_2 - s_1)] \\[1mm] \lambda_1\lambda_2/[s_2\,(s_2 - s_1)] \end{array} \right] e^{s_2 t}
$$

for $t \geqslant 0$, where s_1 and s_2 are the two roots of Λ/s.

The availability $a(t)$ of the system can be obtained from these equations as a function of time. In particular, since the *system availability* $a(t)$ is by definition the probability that at time t the system is in operation, we see that

$$
\begin{aligned}
a(t) &= p_0(t) + p_1(t) \\[2mm]
&= \frac{\mu_2(\mu_1 + \lambda_1)}{s_1 s_2} \\[2mm]
&+ \frac{s_1{}^2 + (\mu_1 + \mu_2 + \lambda_1 + \lambda_2)s_1 + (\mu_1\mu_2 + \lambda_1\mu_2)}{s_1\,(s_1 - s_2)}\, e^{s_1 t} \\[2mm]
&+ \frac{s_2{}^2 + (\mu_1 + \mu_2 + \lambda_1 + \lambda_2)s_2 + (\mu_1\mu_2 + \lambda_1\mu_2)}{s_2\,(s_2 - s_1)}\, e^{s_2 t}
\end{aligned}
$$

and therefore system unavailability is given by

$$
u(t) = 1 - a(t) = p_2(t)
$$

The *reliability* $r(t)$ of a system is defined as the probability that the system has been functioning from the start of operation, time 0, to the present time, time t. To arrive at this measure, we set $\mu_2 = 0$, making $p_2(t)$ an *absorbing state*. Then $r(t)$ is given by the sum of $p_0(t)$ and $p_1(t)$, for $\mu_2 = 0$. That is,

$$
r(t) = \frac{\tilde{s}_1 + (\mu_1 + \lambda_1 + \lambda_2)}{(\tilde{s}_1 - \tilde{s}_2)}\, e^{\tilde{s}_1 t} + \frac{\tilde{s}_2 + (\mu_1 + \lambda_1 + \lambda_2)}{(\tilde{s}_2 - \tilde{s}_1)}\, e^{\tilde{s}_2 t}
$$

where $\tilde{s}_1$ and $\tilde{s}_2$ are the roots of $\Lambda/s\,|_{\mu_2 = 0}$.

Note that the difference between $a(t)$ and $r(t)$ is owing to the fact that we consider repairable systems, since from the above discussion it is clear that $a(t) = r(t)$ for systems with no repairs. Also note that $\lim_{t \to \infty} r(t) = 0$, that is, the probability of no failures tends to zero as t increases, as should be expected. On the other hand, the availability of the system tends to a constant as t increases, in particular,

$$\lim_{t \to \infty} a(t) = \frac{\mu_2(\mu_1 + \lambda_1)}{(\lambda_1\mu_2 + \mu_1\mu_2 + \lambda_1\lambda_2)}$$

With the above assumptions it is also possible to develop distributions for the probability of experiencing n failures as a function of time (the distribution of the number of renewals), as well as the distribution of time of the nth renewal [20].

Example 3.2 **Erlang B Formula** - We now consider a compartmental model that has found extensive use in communication theory, a representation of which is shown in Fig. 3.4.

The differential equations describing this system are given by

$$\dot{p}_0(t) = \mu \, p_1(t) - \lambda \, p_0(t)$$
$$\dot{p}_i(t) = (i+1) \, \mu \, p_{i+1}(t) + \lambda \, p_{i-1}(t)$$
$$\qquad - (\lambda + i\mu) \, p_0(t) \qquad\qquad i = 1,\dots,n$$

The steady-state solution to this system of equations is

$$p_k = p_0 \prod_{i=0}^{k-1} \frac{\lambda}{(i+1)\mu} \qquad k \le n$$

or

$$p_k = p_0 \begin{cases} p_0 (\lambda/\mu)^k 1/k! & k \le n \\ 0 & k > n \end{cases}$$

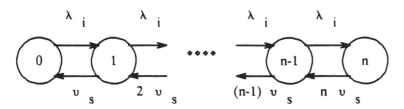

Figure 3.4   State Transition Graph.

and

$$p_0 = \left[\sum_{k=0}^{n} (\lambda/\mu)^k 1/k! \right]^{-1}$$

We see that the probability of being in state p_n, denoted by $B(n,\lambda/\mu)$, is given by

$$p_n \triangleq B(n,\lambda/\mu) = \frac{(\lambda/\mu)^n/n!}{\sum_{k=0}^{n} (\lambda/\mu)^k/k!}$$

The compartmental model of Fig. 3.4 can be used to study trunk group blocking. We consider a trunk group with n trunks where traffic is offered to this trunk group following a Poisson process. $B(n,\lambda/\mu)$ then represents the probability that a call offered to the trunk group will be blocked. This formula was first established by Erlang in 1917 and is known as the Erlang B formula. We will come across it again in Chapter 9.

As already mentioned, our general system of equations (3.9) arose from a compartmental model which restricts the general linear system (3.10) to those which satisfy certain nonnegativity parameter conditions. Results specific to this class of dynamical systems are available; see for instance [2,7,8,13,19]. We now briefly review some of these results, refer to the references for a more detailed discussion.

The basic question that we consider relates to the conditions under which a given system

$$\dot{x}(t) = A x(t) \tag{3.11}$$

is stable. Although, as already discussed, there are various system characteristics of interest, we shall concentrate only on the question of stability. The interested reader is referred to [6,9,15] for general systems and to [7] (and references therein) for compartmental systems. Theorems on controllability and observability of compartmental systems can be found in [11] and [24].

We begin with a definition which will be needed in the sequel. A matrix K is called *reducible* [12] if there exists a nonsingular matrix P such that

$$P K P^{-1} = \begin{pmatrix} K_1 & O \\ K_2 & K_3 \end{pmatrix}$$

where O denotes the zero matrix and K_1 and K_3 are square matrices, all of appropriate dimension. If the matrix K is not reducible, then it is called *irreducible*.

For linear, time-invariant systems, we consider the case in which both A and B are constant matrices. We first note that through a theorem of Gershgoring it can be shown that since $a_{ij} \geqslant 0$ for $i \neq j$, and $a_{ii} \geqslant 0$, then the real parts of all eigenvalues are nonpositive and no pure imaginary roots exist [23]. Thus, the system (3.11), where A is constant, is stable and any oscillations are heavily damped. Furthermore, if the system is closed, that is, a_{0i} and b_i in equations (3.9) are zero for all $i = 1,2,...,n$, then A is singular [13]. This result follows since here the columns of A sum to zero. Furthermore, when a constant input u is applied to (3.3), then the resulting system has no periodic oscillations and has the unique equilibrium point $\hat{x} = -A^{-1}u$. The system is asymptotically stable if and only if $\det A \neq 0$ [16].

Theorem 3.3 [16]   - Consider the matrix $A = (a_{ij})$, with $a_{ij} \geqslant 0$ for $i \neq j$ and $a_{0j} = -\sum_{i=1}^{n} a_{ij} \leqslant 0$. Then $\det A \neq 0$ if and only if there is a path from every compartment to the environment consisting of positive coefficients $a_{ij} > 0$.

A path is defined as a sequence of flows (rates) between compartments where the terminal compartment of a given flow is also the initial compartment of the next flow in the sequence. Thus, all eigenvalues of A have strictly negative real parts if and only if there is a path from every compartment to the environment composed of positive rates. We next state a result which considers the multiplicity of zero eigenvalues of A.

Theorem 3.4 [13]   - Let the system defined by (3.4) be a closed compartmental system. Then if A is an irreducible matrix and $\lambda = 0$ is a root of A, $\lambda = 0$ is a simple root. If $\lambda = 0$ is a multiple root of A, then A is reducible. Clearly if the system is closed, then A has at least one zero eigenvalue since then the elements in any column sum to zero.

As already mentioned, compartmental systems posses no purely imaginary roots. The following result presents conditions under which the system has only real eigenvalues.

Theorem 3.5 [21]   - The eigenvalues of A are real if

$$a_{i_1 i_2} \, a_{i_2 i_3} \, ... \, a_{i_k i_1} = a_{i_2 i_1} \, a_{i_3 i_2} \, ... \, a_{i_1 i_k}$$

for all $i_1, i_2, ... , i_k$ distinct, $k = 1,2,...,n$.

We now consider periodic systems. This is of course the simplest generalization of time-invariant systems. For the next two results we consider systems for which $A(t)$ is piecewise continuous and of period T, and as before, $a_{ij}(t) \geqslant 0$ for $i \neq j$ and for all t. The first result applies to homogeneous systems and states conditions under which a unique solution exists. The second result deals with non-autonomous periodic systems.

Theorem 3.6 [2] - Let the time-varying matrix $A(t)$ be piecewise continuous and periodic of period T. Suppose that there exists a t_0 such that $A(t_0)$ is irreducible, and a vector $v > 0$, $v \in \mathbb{R}^n$ with $A(t)^T v = 0$. Then $x(t)^T v$ is constant along any solution of (3.11), and for a given constant M, there is a unique periodic solution $\hat{x}(t)$ of period T such that $\hat{x}(t)^T v = M$. Every solution $x(t)$ of (3.11), with $x(t)^T v = M$, approaches $\hat{x}(t)$ exponentially when $t \to +\infty$. If $M > 0$, then $\hat{x}(t) > 0$ for all t.

Theorem 3.7 [2] - Let $A(t)$ be a time-varying $n \times n$ matrix, with $A(t_0)$ irreducible for some t_0. Suppose that there exist a $u > 0$ such that $A^T(t)u \leqslant 0$ and $A^T(t_1)u \neq 0$ for some t_1. Let $b(t)$ be a piecewise continuous vector function which is periodic of period T. Then the system $\dot{x} = A(t)x + b(t)$ has a unique periodic solution $x(t)$ of period T to which every other solution approaches exponentially as $t \to \infty$.

The final result of the section considers time-varying, non-periodic systems and presents conditions under which the zero vector is an asymptotically stable equilibrium point.

Theorem 3.8 [2] - Let $A(t)$ be a time-varying $n \times n$ matrix. Suppose that there exist $u > 0$ such that $A^T(t)u \leqslant 0$, and there are i) an infinite family of non-overlapping intervals I_p of width $|I_p|$, $p = 1,2,3,\ldots$, ii) positive constants a and b, iii) an index i_0, so that $|I_p| \geqslant a$ and $[A^T(t)u]_{i_0} \leqslant -b$ for $t \in \bigcup_p I_p$. Furthermore, suppose that there is a constant $c > 0$ such that if i is an index not equal to i_0, $1 \leqslant i \leqslant n$, then there is a sequence of indices $i_1, i_2, \cdots, i_\pi = i$ such that $a_{i_k i_{k+1}}(t) \geqslant c$ for $t \in \bigcup_p I_p$ and $0 \leqslant k \leqslant \pi$. Then $\lim_{t \to \infty} x(t) = 0$ for any solution $x(t)$ of $\dot{x} = A(t)x$.

The conditions in the above theorem basically guarantee that enough "material" flows from the compartments out into the environment and that the compartments are properly connected.

As is usually the case, nonlinear compartmental models are not as well developed as their linear counterparts. The network models with which we deal in Chapter 9, by necessity, belong to the class of nonlinear compartmental models.

3. NONLINEAR COMPARTMENTAL MODELS

The system (3.9) [or (3.10)], as we have seen, can be generalized by letting the a_{ij}'s be time-dependent. However, in both cases it was assumed that the "flow rates" are independent of the concentrations in each compartment, $x_i(t)$. In some systems, such as communication networks in which the amount of call blocking depends on the probing load and carried load, the assumption of linearity is not appropriate. Here the flow rates of each compartment depend on the "concentrations" of several other compartments, leading to a set of nonlinear differential equations of the general form

$$\frac{dx_i}{dt} = b_i(x) + \sum_{\substack{j=1 \\ j \neq i}}^{n} a_{ij}(x)x_j - \sum_{\substack{j=1 \\ j \neq i}}^{n} a_{ji}(x)x_i - a_{0i}(x)x_i \qquad (3.12)$$

Conditions for the existence and uniqueness of the solution of (3.12), as well as the existence of equilibrium solutions, first appeared in [5]. We now present some conditions leading to the existence of equilibrium points for the nonlinear compartmental system, we must however keep in mind that there are many nonequilibrium phenomena associated with open, nonlinear systems [18]. Under certain conditions such systems can exhibit bistability and self-organization.

Suppose that, for each initial condition, Equation (3.12) has a unique solution, and that there exists a positive constant k such that $\sum_{i=1}^{n} (b_i - a_{0i}) \leqslant 0$ for $\sum_{i=1}^{n} x_i \geqslant k$, then it can be shown [19] that (3.12) has an equilibrium point. Furthermore, suppose that for all non-negative real vectors u and w such that $u \leqslant w$ and $u \neq w$, $\sum_{i=1}^{n} [b_i(u) - a_{0i}(u)] \geqslant \sum_{i=1}^{n} [b_i(w) - a_{0i}(w)]$ and $(dx_i/dt)|_{x_i=u_i} \geqslant (dx_i/dt)|_{x_i=w_i}$ for each i. Then the equilibrium point is unique [19].

Theorem 3.9 [19] - Suppose that for all non-negative real vectors u and w such that $u \leqslant w$ and $u \neq w$,

i. $\sum_{i=1}^{n} [b_i(u) - a_{0i}(u)] > \sum_{i=1}^{n} [b_i(w) - a_{0i}(w)]$

ii. $\dfrac{dx_i}{dt}\bigg|_{x_i=u_i} \geqslant \dfrac{dx_i}{dt}\bigg|_{x_i=w_i}$ for each i

Furthermore, suppose that $\sum_{i=1}^{n} [b_i(w) - a_{01}(w)] \rightarrow -\infty$ as $\sum_{i=1}^{n} w_i \rightarrow \infty$,

$(dx/dt)|_{x=0} \geqslant 0$, and that for any non-negative real vector v there exists a constant c such that $|(dx/dt)|_{x=u} - (dx/dt)|_{x=w}| \leqslant c|u-w|$ for $0 \leqslant u, w \leqslant v$. Then for any non-negative initial state x_0, (3.12) has a unique solution and there is a non-negative real vector z such that the solution $x(t) \to z$ as $t \to \infty$.

We now turn to nonlinear, time-invariant systems of the form

$$\frac{dx_i}{dt} = \sum_{j=1}^{n} a_{ij}(x_j) + b_i \tag{3.13}$$

where $a_{ii} = -[a_{0i}(x_i) + \sum_{i \neq j} a_{ji}(x_i)]$ and b_i is constant. We present results concerning (3.13) in terms of the interconnection of compartments and the characteristics of rate functions.

Theorem 3.10 [16] - Let each $a_{ij}(.)$ satisfy a) $a_{ij}(z) \geqslant 0$ for $z \geqslant 0$ and $i \neq j$, b) $a_{ij}(0) = 0$, and let them be differentiable and monotone non-decreasing. Then there is at least one equilibrium point in $\{x|x \geqslant 0\}$ for any constant $b \geqslant 0$ if and only if there is a path from every compartment to the environment that consists of rate functions satisfying $a_{ij}(z) \to +\infty$ as $z \to \infty$.

Theorem 3.11 [16] - Let each $a_{ij}(.)$ be differentiable and monotone non-decreasing. Suppose that (3.13) has a steady-state solution for a non-negative constant input b. Then the steady state is independent of the initial state and uniquely determined by b if each compartment possesses a path to the environment consisting of strictly monotone increasing rate functions.

Theorem 3.12 [16] - Let each a_{ij} satisfy the conditions of Theorem 3.10, and $b = 0$. Then the origin $x = 0$ is globally asymptotically stable with respect to the positive orthant if each compartment possesses a path to the environment consisting of positive rate functions $a_{ij}(z) > 0$ for all $z > 0$.

Similar to the compartmental models described by the system of differential Equations (3.8) and (3.11), one can consider compartmental models described by difference equations. A general compartmental model of this type can be described by the set of difference equations

$$x(k+1) = b_i + \sum_{\substack{j=1 \\ j \neq i}}^{n} a_{ij}[x(k)] \, x_j(k) - \sum_{\substack{j=1 \\ j \neq i}}^{n} a_{ij}[x(k)] \, x_i(k)$$

$$- a_{0i}[x(k)] \, x_i(k)$$

In the study of Chapter 9 we will use both the differential and difference equation formulations of compartmental models. Of course, even the system of differential equations must eventually be transformed into a set of difference equations for implementation on a digital computer.

4. STOCHASTIC COMPARTMENTAL MODELS

So far in the discussion on compartmental models we have concentrated on deterministic models. We now briefly mention corresponding stochastic models. For such systems the notion of state, used in the above formulations, must be slightly modified. In loose terms, the state of a deterministic system consists of the information needed to be able to predict future system behavior. For systems displaying stochastic properties, the state of the system can be defined as the information needed to uniquely determine the probability distributions of the state variables at future times [3]. To maintain this notion of state (i.e., prediction of future values based on present values) we require that the system be modeled as a Markov process. For systems of the form

$$\dot{x}(t) = f(x,t) + g(x,u,t) \tag{3.14}$$

$$y(t) = h(x,t)$$

(as well as their difference equation equivalent), a more general version of (3.9), the corresponding stochastic system representation can be written as

$$\dot{x}(t) = f(x,t) + g(x,u,t) + \sigma(x,t) \, \nu(t) \tag{3.15}$$

where the functions f and g have the same meaning as in (3.14), and the last term represents the "noise" added to the system with ν a random disturbance function. To maintain our notion of state, $\nu(t)$ is considered to be a continuous-time white noise process, or a process that can be generated from white noise.†
The corresponding stochastic differential equation is

$$dx = f(x,t) \, dt + g(x,u,t) \, dt + \sigma(x,t) \, d\omega \tag{3.16}$$

---

†Note that stochastic processes, with covariance functions belonging to a large class of processes, can be obtained by passing white noise through an appropriate linear system [3].

where $d\omega$ denotes the increment of a Wiener process. Equation (3.14) can be equivalently written as

$$x(t) = x(t_0) + \int_{t_0}^{t} f[x(s),s] \, ds + \int_{t_0}^{t} g[x(s),u(s),s] \, ds \qquad (3.17)$$
$$+ \int_{t_0}^{t} \sigma[x(s),s] \, d\omega(s)$$

where the last integral can be considered to be an Ito integral.

With this formulation, the basic problem then consists of estimating the system's state based on "noisy" measurements which can be accomplished using appropriate smoothing, filtering or prediction techniques depending on whether we are trying to obtain past, present, or future (respectively) values of the state variables. Refer to [3] for a complete discussion of such systems, and to [22] for stochastic compartmental systems in particular.

5. APPLICATION OF COMPARTMENTAL MODELS TO COMMUNICATIONS NETWORKS

Thus far we have discussed "substances," material transfer, and compartments which are certainly not part of the standard terminology for communications networks. However, the formalism described fits in naturally with the modeling of such networks. At any one time, we can consider our network as consisting of various populations of calls. Examples of such populations are the set of currently connected calls (carried load) and the set of probing calls. With this setup, the substances refer to the various populations, a compartment is made up of a single population of calls, and material transfer among compartments refers to a change in status of a given call, which causes a change in the population that the call belongs to. For example, the change in status when a probing call finally reaches and connects to its destination. When considering the two populations in a network, probing calls and connected calls, the inherent nonlinearity of the system becomes clear. That is, the percentage of probing calls that connect to their destination is not only a function of the number of calls (concentration) in the probing state, but also the number of calls in the connected state (carried load) since all these calls use resources in the network. Although the use of compartmental analysis in engineering has been recognized, very little work has actually been done in this area. Chapter 9 presents a natural application of compartmental analysis to the study of communications networks.

REFERENCES

[1] Anderson, B. D. O. and Parks, P. C. - "Lumped Approximations of Distributed Systems and Controllability Questions," *Proceedings of the IEEE* 132 (D) (3), May 1985.

[2] Aronsson, G. and Kellogg, R. B. - "On a Differential Equation Arising from Compartmental Analysis," *Mathematical Biosciences* 38, pp. 113-122, 1978.

[3] Astrom, K. J. - *Introduction to Stochastic Control Theory*, Academic Press, 1970.

[4] Barnett, S. and Cameron, R. G. - *Introduction to Mathematical Control Theory*, Clarendon Press, 1985.

[5] Bellman, R. - "Topics in Pharmacokinetics I: Concentration-Dependent Rates," *Mathematical Biosciences* 6, pp. 13-17, 1970.

[6] Brockett, R. W. - *Finite Dimensional Linear Systems*, Wiley, 1969.

[7] Brown, R. F. - "Compartmental System Analysis: State of the Art." *IEEE Transactions on Biomedical Engineering* BME-27 (1), pp. 1-11, January 1980.

[8] Carson, E. E., Cobelli, C., and Finkelstein, L. - *The Mathematical Modeling of Metabolic and Endocrine Systems*, Wiley, 1983.

[9] Casti, J. L. - *Dynamical Systems and their Applications: Linear Theory*, Academic Press, 1977.

[10] Casti, J. L. - *Nonlinear System Theory*, Academic Press, 1984.

[11] Cobelli, C. and Romanin-Jacur, G. - "Controllability, Observability and Structural Identifiability of Multi Input and Multi Output Biological Compartmental Systems." *IEEE Transactions on Biomedical Engineering* BME-23 (2), March 1976.

[12] Gantmacher, F. R. - *The Theory of Matrices* - Vols. I & II. Chelsea, 1959.

[13] Hearon, J. Z. - "Theorems on Linear Systems," *Annals New York Academy of Sciences* 108, pp. 36-91, 1963.

[14] Hirsch, M. W. and Smale, S. - *Differential Equations, Dynamical Systems, and Linear Algebra*, Academic Press, 1974.

[15] Kalman, R. E., Falb, P. L., and Arbib, M. A. - *Topics in Mathematical Systems Theory*, McGraw-Hill, 1962.

[16] Maeda, H. and Kodama, S. - "Asymptotic Behavior of Nonlinear Compartmental Systems: Nonoscillation and Stability," *IEEE Transactions on Circuits and Systems* CAS-25 (6), pp. 372-378, 1978.

[17] Kleinrock, L. - *Queuing Systems, Volume I: Theory*, Wiley, 1975.

[18] Prigogine, I. and Nicolis, G. - *Self-Organization in Nonequilibrium Systems*, Wiley, 1977.

[19] Sandberg, I. W. - "On The Mathematical Foundations of Compartmental Analysis in Biology, Medicine, and Ecology," *IEEE Transactions on Circuit and Systems* CAS-25 (5), pp. 273-279, 1978.

[20] Shooman, M. L. - *Probabilistic Reliability: An Engineering Approach*, McGraw-Hill, 1968.

[21] Smith, W. D. and Mohler, R. R. - "Necessary and Sufficient Conditions in the Tracer Determination of Compartmental System Order," *Journal of Theoretical Biology* 57, pp. 1-21, 1976.

[22] Thakur, A. K., Rescigno, A., and Schafer, D. E. - "On the Stochastic Theory of Compartments: II. Multi-Compartment Systems," *Bulletin of Mathematical Biology* 35, 1973.

[23] Walter, G. G. - "Complexity and Stability in Compartmental Models," *SIAM Journal on Algebraic and Discrete Methods* 6 (1), pp. 39-46, January 1985.

[24] Zazworsky, R. M. and Knudsen, H. K. - "Controllability and Observability of Linear Time-Invariant Compartmental Models," *IEEE Transactions on Automatic Control* AC-23 (5), 1978.

II
Computer Networks

4

Birth-Death Processes

Ricardo F. Garzia

AT&T Bell Laboratories
Columbus, Ohio

1. INTRODUCTION

The intuitive idea behind the birth-death process may be better understood if we consider a *population* that is simultaneously gaining new members through a birth process and loosing old members through a death process, as is the case with, say, the human population on earth. In a computer system, we have another example of this type of process. Transactions arrive at the CPU for processing, these transaction arrivals correspond to births. Transactions leave the CPU when they are finished processing, transaction departure corresponds to a death. Any system that has only births or deaths is called a pure system. Then, we can have a pure-birth process or a pure-death process; see [1-5].

We now make some definitions: the *state* of a network is defined as the number of transactions currently in the network and is designated as S_n for a network state with a population of n. The succession of network state changes is called a Markovian chain, i.e., the states shown in Figure 4.1, $(S_1, S_2, S_3, \cdots, S_n)$, form a Markovian chain.

The birth-death process is a special case of a Markov process in which transactions from one state S_n are permitted only to neighboring states S_{n+1}, S_n, S_{n-1}. This is a strong restriction that often allows us to arrive at

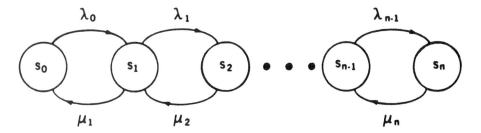

Figure 4.1   State Transitions.

a closed form solution. Therefore, given a state S_n at time t, the state at $(t + \Delta t)$ falls into one of the following cases:

- We were in state S_n at time t and no state change occurred.
- We were in state S_{n-1} at time t and at $(t + \Delta t)$ we changed to state S_n; therefore, one birth occurred.
- We were in state S_{n+1} at time t, and by time $(t + \Delta t)$ one death occurred.

In Fig. 4.1 we show $n + 1$ states and the corresponding relations. Note that each state is related to neighboring states; for example, we may move to state S_i only from state S_{i-1} or state S_{i+1}.

If we take into consideration the time involved, which could be discrete or continuous, the birth-death process can be categorized as a

- Continuous-Time Process
- Discrete-Time Process

The discrete-time birth-death process is of less importance to us than the continuous-time birth-death process, since we are interested in processes that can be represented by a Poisson distribution. Therefore, our main interest will focus on continuous-time birth-death processes with a discrete state space in which transitions only move to neighboring states S_{n+1} or S_{n-1} from a given state S_n.

A birth-death process is appropriate for modeling changes in the size of a population. Indeed, the process is said to be in state S_n when the population at that time is of size n. Thus,

$$S_n \rightarrow S_{n+1} \qquad birth$$

$$S_n \rightarrow S_{n-1} \qquad death$$

We consider changes in the size of a population where transitions result in an increase or decrease, only one member, in the population size. Let

$$\lambda_n \; - \; average \; birth \; rate$$

$$\mu_n \; - \; average \; death \; rate$$

In this model, we assume that these two average rates are not a function of time, and they are only defined by the population size n. They are called a *continuous-time homogeneous Markov chain of the birth-death type*.

Now we are interested in the differential-difference equation that allows us to establish the probability of being in state S_n at time t, denoted by

$$P_n(t) = P[\xi(t)=n] \tag{4.1}$$

If $n \geqslant 1$, the probability $P_n(t+\Delta t)$ that the system will be in state S_n at time $t+\Delta t$ has three components

$$P_n(t) = P_{n1}(t) + P_{n2}(t) + P_{n3}(t) \tag{4.2}$$

In what follows we will use the notation $o(\Delta t)$, where a function $f(\Delta t)$ is said to be $o(\Delta t)$ if

$$\lim_{\Delta t \to 0} \frac{f(\Delta t)}{\Delta t} = 0$$

If we assume an exponential distribution

$$P[\xi \leqslant \Delta t] = 1 - e^{-\lambda \Delta t}$$

the exponential part of the above equation can be expanded as a series

$$e^{-\lambda \Delta t} = 1 - \lambda \Delta t + \frac{(\Delta t)^2}{2} - \frac{(\lambda \Delta t)^3}{6} + \dots.$$

$$= 1 - \lambda \Delta t + o(\Delta t)$$

To obtain the value for each component of Equation (4.2) we consider:

1. The probability that the birth-death process was in state S_n at time t and no transitions occurred, either births or deaths, implies

$S_n \rightarrow S_{n+1}$ *did not occurr; thus,* $P_{n11}(t+\Delta t) = 1 - \lambda_n \Delta t + o_1(\Delta t)$

$S_n \rightarrow S_{n-1}$ *did not occurr; thus,* $P_{n12}(t+\Delta t) = 1 - \mu_n \Delta t + o_2(\Delta t)$

Therefore, the first component is given by

$$P_{n1}(t+\Delta t) = P_n(t) \ [1 - \lambda_n \Delta t + o_1(\Delta t)]$$
$$\times \ [1 - \mu_n \Delta t + o_2(\Delta t)] \qquad (4.3)$$

Expanding Equation (4.3), we can write

$$P_{n1}(t+\Delta t) = P_n(t) \ [1 - \lambda_n \Delta t - \mu_n \Delta t + o_1(\Delta t)$$
$$+ o_2(\Delta t) + \lambda_n \mu_n \Delta t^2 - \lambda_n \ \Delta t \ o_2(\Delta t)$$
$$- \mu_n \ \Delta t \ o_1(\Delta t) + o_1(\Delta t) o_2(\Delta t)]$$

which can be written as

$$P_n(t+\Delta t) = P_n(t) \ [1 - \lambda_n \Delta t - \mu_n \Delta t + o(\Delta t)] \qquad (4.4)$$

where $o(\Delta t)$ now incorporates the following terms

$$o(\Delta t) = o_1(\Delta t) + o_2(\Delta t) - \mu_n \ \Delta t \ o_1(\Delta t)$$
$$- \lambda_n \ \Delta t \ o_2(\Delta t) + o_1(\Delta t) o_2(\Delta t) \qquad (4.5)$$

2. The probability that the system was in state S_{n-1} at time t, times the probability that the transition $S_{n-1} \rightarrow S_n$ occurred in the interval of time Δt, leads us to the second component of the probability

$$P_{n2}(t+\Delta t) = P_{n-1}(t) \ [\lambda_{n-1} \Delta t + o_3(\Delta t)] \qquad (4.6)$$

3. The probability that the system was in state S_{n+1} at time t, multiplied by the probability that the transition $S_{n+1} \rightarrow S_n$ occurred, results in the third component

$$P_{n3}(t+\Delta t) = P_{n+1}(t) \ \mu_{n+1} \ \Delta t + o_4(\Delta t) \qquad (4.7)$$

Combining all previous equations, we have

$$P_n(t+\Delta t) = [1 - \lambda_n \Delta t - \mu_n \Delta t] \ P_n(t) + \lambda_{n-1} \ \Delta t \ P_{n-1}(t)$$
$$+ \mu_{n+1} \ \Delta t \ P_{n+1}(t) + o(\Delta t) \qquad (4.8)$$

The solution of Equation (4.8) is the probability that the network is in a given state at a given time, which defines the dynamics of the network. To find the solution of Equation (4.8) we may go from a regular application of calculus to a more sophisticated mixed Laplace and z transform approach. Whichever approach is taken, it is more convenient to rearrange Equation (4.8) as follows

$$\frac{P_n(t+\Delta t) - P_n(t)}{\Delta t} = -(\lambda_n + \mu_n)P_n + \lambda_{n-1}P_{n-1}(t)$$

$$+ \mu_{n+1}P_{n+1}(t) + \frac{o(\Delta t)}{\Delta t} \qquad (4.9)$$

Taking the limit of Equation (4.10) when $\Delta t \to 0$, we obtain

$$\frac{dP_n(t)}{dt} = -(\lambda_n + \mu_n) P_n(t) + \lambda_{n-1}P_{n-1}(t) + \mu_{n+1}P_{n+1}(t) \qquad (4.10)$$

The above equation is valid for $n \geqslant 1$. For a solution we need to consider the initial conditions.

For n = 0, we get

$$\frac{dP_0(t)}{dt} = -\lambda_0 P_0(t) + \mu_1 P_1(t) \qquad (4.11)$$

The birth-death process is defined by the infinite set of differential-difference equations with initial conditions. It can be shown that this set of differential-difference equations has a solution $P_n(t)$ for all n and t under very general conditions. However, the solution is difficult to obtain analytically except for some special cases.

Let us now go back to the birth-death rates given by λ_n and μ_n. In accordance with Kleinrock's notation [4], we write these rates as follows

$$\lambda_n = q_{n,n+1} \qquad (4.12)$$

$$\mu_n = q_{n,n-1} \qquad (4.13)$$

The notation introduced by Kleinrock leads directly to the queuing network notation.

The nearest neighbor condition requires that all $q_{kj} = 0$ for $|k-j| > 1$. Also

$$\sum q_{kj} = 0 \qquad (4.14)$$

Therefore, we have

$$q_{kk} = -(\mu_k + \lambda_k) \qquad (4.15)$$

Thus, the infinitesimal generator for our system is

$$Q = \begin{bmatrix} -\lambda_0 & \lambda_0 & 0 & 0 & 0 & \cdots \\ \mu_1 & -(\lambda_1+\mu_1) & \lambda_1 & 0 & 0 & \cdots \\ 0 & \mu_2 & -(\lambda_2+\mu_2) & \lambda_2 & 0 & \cdots \\ 0 & 0 & \mu_3 & -(\lambda_3+\mu_3) & \lambda_3 & \cdots \\ \cdots & \cdots & \cdots & \cdots & \cdots & \cdots \end{bmatrix} \qquad (4.16)$$

We note in Equation (4.16) that except for the main, upper and lower diagonals, all terms are zero.

2.   BIRTH-DEATH SYSTEMS IN EQUILIBRIUM

We have discussed some general concepts of the birth-death process. We now concentrate on birth-death systems in equilibrium; see [4].

2.1   General Equilibrium Solution

For our present studies we are interested in obtaining probabilities that are not functions of time. Let us define the probability given by

$$p_n = \lim_{t \to \infty} P_n(t) \qquad (4.17)$$

where the probability p_n represents the probability of having n members in our system for an arbitrary amount of time in the distant future.

In the previous definition the word *arbitrary* is of paramount importance, because it implies the independence from time of p_n. We are not claiming that the process does not move from state to state in this limiting case; certainly the number of members in the population will change with time, but the *long-run* probability of finding the system with n members will be properly described by p_n.

Accepting the existence of the limit in Equation (4.17), we may set

$$\lim_{t \to \infty} \frac{dP_n(t)}{dt} = 0 \qquad (4.18)$$

thereby obtaining the result

$$(\lambda_n+\mu_n)p_n + \lambda_{n-1}p_{n-1} + \mu_{n+1}p_{n+1} = 0 \qquad n \geqslant 1 \qquad (4.19)$$

$$-\lambda_0 p_0 + \mu_1 p_1 = 0 \qquad n = 0 \qquad (4.20)$$

We now reduce the two Equations (4.19) and (4.20) to only one equation. For this purpose we let

$$\lambda_{-1} = \lambda_{-2} = \lambda_{-3} = \cdots = 0 \tag{4.21}$$

$$\mu_0 = \mu_{-1} = \mu_{-2} = \cdots = 0 \tag{4.22}$$

Since we cannot have a negative population size, we set

$$p_{-1} = p_{-2} = p_{-3} = \cdots = 0 \tag{4.23}$$

With the above assumptions, we can combine Equations (4.19) and (4.20) into the following equation

$$- (\lambda_n + \mu_n)p_n + \lambda_{n-1}p_{n-1} + \mu_{n+1}p_{n+1} = 0 \tag{4.24}$$

This equation is valid for all values $n = 0,1,2,3,\ldots$.
The conservation relation is

$$\sum_{n=0}^{\infty} p_n = 1 \tag{4.25}$$

In the equilibrium case it is clear that the flow must be conserved in the sense that for each state the input flow must be equal to the output flow.
Observing Fig. 4.2, we can write

$$Flow\ rate\ into\ S_n = \lambda_{n-1}p_{n-1} + \mu_{n+1}p_{n+1} \tag{4.26}$$

$$Flow\ rate\ out\ of\ S_n = (\lambda_n + \mu_n)p_n \tag{4.27}$$

If the system is in equilibrium, the two flow rates must be equal; thus, we write

$$\lambda_{n-1}p_{n-1} + \mu_{n+1}p_{n+1} = (\lambda_n + \mu_n)p_n \tag{4.28}$$

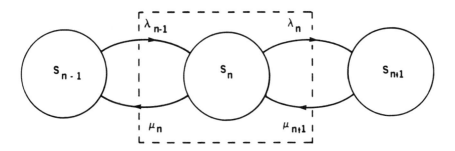

Figure 4.2   State Transition Rate Diagram.

Equation (4.28) is the equilibrium equation for our system. But we are interested in having a simpler equation of equilibrium than the one just obtained. For this purpose, we will define the surrounding boundaries as shown in Fig. 4.3.

From Figure 4.3 we see that we define boundaries for a number of states. The first boundary has only one state (S_0); the second boundary has two states (S_0 and S_1); and so on. The sequence of boundaries is

1st boundary S_0
2nd boundary S_0 and S_1
3rd boundary S_0, S_1, and S_2
....
nth boundary S_0, S_1, $\cdots$, S_{n-1}

The last boundary would lead to the following simple conservation of flow relationship

$$\lambda_{n-1}p_{n-1} = \mu_n p_n \tag{4.29}$$

For the general equilibrium solution we are interested in finding the function that relates p_0 to p_n, or

$$p_n = f(p_0) \tag{4.30}$$

Taking Equation (4.24), we can write, assuming $n=0$,

$$-\lambda_0 p_0 + \mu_1 p_1 = 0 \tag{4.31}$$

Then,

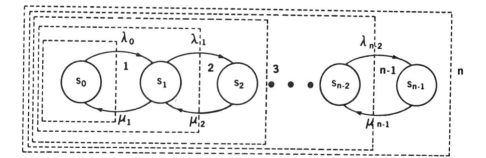

Figure 4.3   State Surrounding Boundaries.

$$p_1 = \frac{\lambda_0}{\mu_1} p_0 \qquad (4.32)$$

Now, let us assume that $n = 1$; using again the same procedure, we can write

$$-(\lambda_1 + \mu_1) + \lambda_0 p_0 + \mu_2 p_2 = 0 \qquad (4.33)$$

Substituting for p_1

$$-(\lambda_1 + \mu_1) \frac{\lambda_0}{\mu_1} p_0 + \lambda_0 p_0 + \mu_2 p_2 = 0 \qquad (4.34)$$

we have

$$p_2 = \frac{\lambda_0 \lambda_1}{\mu_1 \mu_2} p_0 \qquad (4.35)$$

And continuing in this way, we obtain

$$p_n = \frac{\lambda_0 \lambda_1 \lambda_2 \cdots \lambda_{n-1}}{\mu_1 \mu_2 \mu_3 \cdots \mu_n} p_0 \qquad (4.36)$$

which can be written as follows

$$p_n = p_0 \prod_{i=0}^{n-1} \frac{\lambda_i}{\mu_{i+1}} \qquad (4.37)$$

From Equation (4.25), we can write

$$\sum_{n=0}^{\infty} p_n = p_0 + p_0 \sum_{n=1}^{\infty} \prod_{i=0}^{n-1} \frac{\lambda_i}{\mu_{i+1}} \qquad (4.38)$$

or

$$p_0 = \frac{1}{1 + \sum_{n=1}^{\infty} \prod_{i=0}^{n-1} \frac{\lambda_i}{\mu_{i+1}}} \qquad (4.39)$$

Knowing all the birth-death rates, Equation (4.39) allows us to get the value of p_0. Thereafter, Equation (4.37) allows the determination of p_n for all $n \geqslant 1$.

REFERENCES

[1] Allen, A. O. - *Probability, Statistics, and Queuing Theory with Computer Applications*, Academic Press, 1978.

[2] Cooper, R. B. - *Introduction to Queuing Theory* - Second Edition, Elsevier North Holland, 1981.
[3] Gordon, W. J. and Nowell, G. P. - "Closed Queuing Systems with Exponential Servers," *Operations Research* 15, pp. 254-265, 1967.
[4] Kleinrock, L. - *Queueing Systems* - Volumes I and Volume II. Wiley, 1975.
[5] Stuck, B. W. and Arthurs, E. - *A Computer and Communications Network Performance Analysis Primer*, Prentice-Hall, 1985.

5

Birth-Death Queuing Systems

Ricardo F. Garzia

AT&T Bell Laboratories
Columbus, Ohio

1. INTRODUCTION

In previous chapters we presented two different types of basic processes, namely,

- Queuing Processes - Processes that are characterized by a unit providing service, therefore creating, in general, a queue for incoming transactions, which need to wait to obtain service. The mathematical description of these processes was done based on the incoming rate (λ), and the processing rate (μ). These two rates define the mathematical process.

- Birth-Death Processes - These processes are characterized by the birth rate of the population (λ), and the death rate of the population (μ). Since the size of the population changes, we introduce the concept of state. These two rates define the mathematical process which is depicted in Fig. 5.1.

We can see that the two types of processes described are characterized by the same mathematical descriptors. Therefore, we are going to study the basic birth-death queuing system in equilibrium. After this presentation we will move to Markovian Queues in Equilibrium, a more general birth-death process.

In order to obtain the state probabilities in equilibrium, we use Equations (4.37) and (4.39) which are repeated below for convenience.

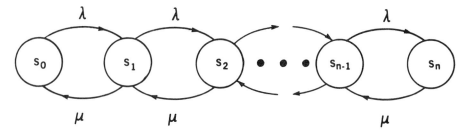

Figure 5.1   State-Transition-Rate Diagram for M/M/1.

$$p_n = p_0 \prod_{i=0}^{n-1} \frac{\lambda_i}{\mu_{i+1}} \tag{5.1}$$

$$p_0 = \frac{1}{1 + \sum_{n=1}^{\infty} \prod_{i=0}^{n-1} \frac{\lambda_i}{\mu_{i+1}}} \tag{5.2}$$

We will describe in this chapter those queuing systems that are of interest to us. To define these systems we use the *Kendall* notation. This notation uses the interarrival time distribution, service time distribution, and number of servers, as follows:

interarrival time distribution/service time distribution/number of servers

The letter M (Markov) means exponential. The systems to be described are:

• M/M/1 - Classical Queuing System
• M/M/∞ - Responsive Servers

This chapter represents a selection of topics extracted from [1]. Refer to [1] for a more complete discussion.

1.1   M/M/1 - Classical Queuing System

The model assumes a random (Poisson) arrival pattern and a random (exponential) service time distribution. The birth and death rates do not depend upon the number of customers or transactions in the system. Therefore,

$$\lambda_0 = \lambda_1 = \lambda_2 = \ldots = \lambda \tag{5.3}$$

$$\mu_1 = \mu_2 = \mu_3 = \ ..... \ = \mu \qquad (5.4)$$

We further assume that infinite queue space is provided, and that we have only one server.

The probability p_n is given by

$$p_n = p_0 \prod_{i=0}^{n-1} \frac{\lambda}{\mu} = p_0 (\frac{\lambda}{\mu})^n \qquad n \geqslant 0 \qquad (5.5)$$

And, therefore, the value of p_0 is given by

$$p_0 = \frac{1}{1 + \sum_{n=0}^{\infty} \prod_{i=0}^{n} \frac{\lambda}{\mu}} = \frac{1}{1 + \sum_{n=1}^{\infty} (\frac{\lambda}{\mu})^n} = \frac{1}{\sum_{n=0}^{\infty} (\frac{\lambda}{\mu})^n}$$

$$= \frac{1}{\dfrac{1}{1 - \dfrac{\lambda}{\mu}}} = 1 - \frac{\lambda}{\mu} \qquad (5.6)$$

The sum converges if $\lambda < \mu$. Therefore,

$$\sum_{n=1}^{\infty} (\frac{\lambda}{\mu})^n = \frac{\lambda}{\mu} + (\frac{\lambda}{\mu})^2 + (\frac{\lambda}{\mu})^3 + (\frac{\lambda}{\mu})^4 + \ .....$$

$$= \frac{\lambda}{\mu} \left[1 + \frac{\lambda}{\mu} + (\frac{\lambda}{\mu})^2 + (\frac{\lambda}{\mu})^3 + \ ..... \right] \qquad (5.7)$$

$$= \frac{\lambda}{\mu} \frac{1}{1 - \dfrac{\lambda}{\mu}}$$

Then, Equation (5.7) becomes

$$p_0 = \frac{1}{1 + \dfrac{\lambda/\mu}{1 - \dfrac{\lambda}{\mu}}} = 1 - \frac{\lambda}{\mu} \qquad (5.8)$$

Let us define ρ as the ratio of λ and μ

$$\rho = \frac{\lambda}{\mu} \qquad (5.9)$$

Then, from our stability condition, we require that $0 \leqslant \rho < 1$; this insures that $p_0 > 0$. Then

$$p_n = (1 - \rho) \rho^n \tag{5.10}$$

We make the important observation that p_n depends on λ and μ only through the ratio ρ. The above equation is indeed the solution for the steady-state probability of having n customers in the system.

An important measure of a queuing system is the average number of customers in the system $\bar{N}$.

$$\bar{N} = \sum_{n=0}^{\infty} n\, p_n = (1 - \rho) \sum_{n=0}^{\infty} n\, \rho^n = \frac{\rho}{1 - \rho} \tag{5.11}$$

The average time spent in the system by a customer is given by

$$T = \frac{\bar{N}}{\lambda} = \frac{1/\mu}{1 - \rho} \tag{5.12}$$

The dependence of the average time on the utilization factor (ρ) is shown in Fig. 5.2.

The standard deviation is given by

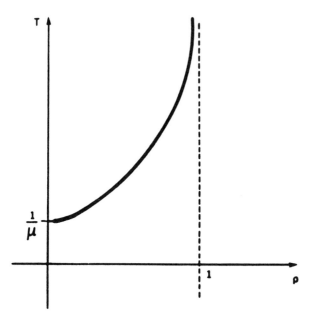

Figure 5.2   Average Delay as a Function of ρ for M/M/1.

$$\sigma_N^2 = \sum_{n=0}^{\infty} (n - \bar{N})^2 \, p_n = \frac{\rho}{(1 - \rho)^2} \qquad (5.13)$$

Example 5.1 - Let us assume that we have a small batch computing system as shown in Fig. 5.3. We further assume that the I/O device can process almost instantaneously the transactions that finish processing at the CPU. Therefore, there is only one queue in our system. The jobs arrive at the input queue every 4 minutes. These jobs are processed on a first-come-first-served basis. The processing time of each transaction is assumed to be exponentially distributed with an average time of 3 minutes.

We are interested in the information that follows:

1. What is the probability that an arriving job will require more than 20 minutes to be processed?
2. A queue of jobs waiting to be processed will form occasionally. What is the average number of jobs waiting in the queue?
3. It is decided that when the work load increases to the level such that the average time in the system reaches 30 minutes, the computer system capacity will be increased. What is the average arrival rate of jobs per hour at which this will occur? What is the percentage increase over the present job load ? What is the average number of jobs in the system at this time ?
4. Suppose the criterion for upgrading the computer capacity is that not more than 10 percent of all jobs have a time in the system exceeding 40 minutes. At the arrival rate for which this criterion is reached, what is the average number of jobs waiting to be processed?

The response to the first question follows.
We know that

$$\rho = \frac{\lambda}{\mu}$$

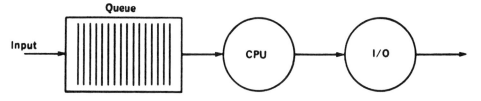

Figure 5.3 Small Batch Computing System.

Therefore,

$$\rho = \frac{1/4}{1/3} = 0.75$$

The average time in the system is given by

$$T = \frac{1/\mu}{1 - \rho} = \frac{3}{1 - 0.75} = 12 \ minutes$$

The probability that an arriving job will require more that 20 minutes is given by

$$P[w>t] = 1 - e^{-t/12}$$

which can be written as

$$P[w>t] = e^{-t/12} = e^{-20/12} = 0.1889$$

The response to the second question is

$$\bar{N} = \frac{\rho}{1 - \rho} = \frac{0.75}{1 - 0.75} = 3 \ jobs$$

The response to the third question follows

$$T = \frac{1/\mu}{1 - \rho} = \frac{1/\mu}{1 - \lambda \ 1/\mu}$$

In our case

$$T = 30 = \frac{3}{1 - 3\lambda}$$

which results in $\lambda = 0.3$. The present increase in the work load is

$$For \ \lambda = \frac{1}{4} = 0.25 \ represents \ 15 \ jobs/hr$$

$$For \ \lambda = 0.3 \ represents \ 54 \ jobs/hr$$

Therefore, the load increase is

$$\frac{54 - 15}{15} = 2.6$$

The average number of jobs is

$$\bar{N} = \frac{\rho}{1 - \rho} = \frac{0.9}{1 - 0.9} = 9$$

1.2 M/M/∞

We now present a system that can be interpreted in two different ways:
1. A responsive server system which accelerates the service rate linearly when more customers are waiting; 2. A system in which there is always a new clerk or server available for each arriving customer.

In particular we set

$$\lambda_n = \lambda \tag{5.13}$$

$$\mu_n = n\mu \tag{5.14}$$

With the birth-death rates given by Equations (5.13) and (5.14) we arrive at the state-transition diagram shown in Fig. 5.4.

The probability p_n is given by

$$p_n = p_0 \prod_{i=0}^{n-1} \frac{\lambda}{(i+1)\mu} \tag{5.15}$$

which can be reduced to

$$p_n = p_0 \frac{(\lambda/\mu)^n}{n!}$$

substituting p_0 from Equation (5.2), we obtain

$$p_n = \frac{(\lambda/\mu)^n}{n!} e^{-\lambda/\mu} \tag{5.16}$$

We are also interested in the average value of the population size, which is given by

$$\bar{N} = \frac{\lambda}{\mu} \tag{5.17}$$

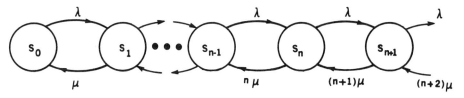

Figure 5.4 State-Transition-Rate Diagram for M/M/∞.

When we use Little's result the above equation becomes

$$T = \frac{1}{\mu}$$
(5.18)

From inspection, we expect the above result, since we provide all the servers needed. Thus, the time spent in the system is the time required for service, as seen in Equation (5.18).

2. MARKOVIAN QUEUES IN EQUILIBRIUM

We now consider a more general Markov chain whose states interact with other non-neighboring states. In this case we can still use the methodology already learned; for this purpose let us consider the system shown in Fig. 5.5.

The equilibrium condition for state S_0 is

$$\left[\lambda + \frac{\lambda}{2} \right] p_0 = \frac{3}{2} \lambda \, p_0 = \mu \, p_1$$
(5.19)

For state S_1 we can write

$$(\lambda + \mu) \, p_1 = \lambda \, p_0 + \mu \, p_2$$
(5.20)

Finally, for state S_2 we have

$$\mu \, p_2 = \frac{\lambda}{2} \, p_0 + \lambda \, p_1$$
(5.21)

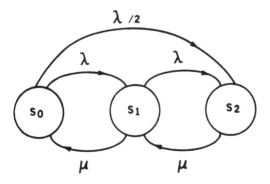

Figure 5.5   Basic Non-Nearest Neighbor System.

Adding $\lambda\ p_0$ to both sides of the equation, we have

$$\lambda\ p_0 + \mu\ p_2 = \frac{3}{2}\lambda\ p_0 + \lambda\ p_1 \tag{5.22}$$

If we use the information provided by Equation (5.22), Equation (5.21) becomes

$$(\mu + \lambda)\ p_1 = \lambda\ p_0 + \mu\ p_2 \tag{5.23}$$

which is Equation (5.20). We always have a redundant equation in finite Markov chains. Therefore, for n states we need only write the $n-1$ state conditions in equilibrium. Also, we know that

$$\sum_{i=0}^{\infty} p_i = 0 \tag{5.24}$$

Therefore in our case

$$p_0 + p_1 + p_2 = 0 \tag{5.25}$$

The solution of the previous equations then becomes

$$p_0 = \left[1 + 2\frac{\lambda}{\mu} + \frac{3}{2}\left(\frac{\lambda}{\mu}\right)^2 \right]^{-1} \tag{5.26}$$

$$p_1 = \frac{3}{2}\frac{\lambda}{\mu}p_0 \tag{5.27}$$

$$p_2 = \left[\frac{1}{2}\frac{\lambda}{\mu} + \frac{3}{2}\left(\frac{\lambda}{\mu}\right)^2 \right] p_0$$

REFERENCE

[1]   Kleinrock, L. - *Queueing Systems* - Volumes I and II, Wiley, 1975.

6

Queuing Network Models

Ricardo F. Garzia

AT&T Bell Laboratories
Columbus, Ohio

1. INTRODUCTION

In this chapter we concentrate on the development of computer network models for network performance evaluation. Although we concentrate on computer network models, similar discussions can be applied to manufacturing networks. These two applications display strong similarities, but in general, the first one is more complex due to the multiprogramming feature. We now introduce some definitions:

1. **Transaction** - The flow of events in a network defined as transactions.
2. **Multiprogramming Level** - Is defined as the number of transactions, of a given class, that are allowed to share processing at a node. Since our model is based on average values, the multiprogramming level is represented by a real number.
3. **Class Hopping** - Refers to transactions that change class after completing processing at a network node.
4. **Queue** - Is the *waiting room* where transactions stay until they begin processing at the node.
5. **Queue Discipline** - Is the queue policy that waiting transactions follow to receive service. Several types of disciplines are in use, including

FIFO (First-In-First-Out), LIFO (Last-In-First-Out), and IS (Infinite Server).

6. **Node Distribution** - Is the continuous random variable distribution representing the processing time. Several distributions are in use including the Poisson distribution, Gamma distribution, and Erlang-k distribution.

We now discuss in some detail the basic queuing system shown in Fig. 6.1. Transactions come into the system in a given time sequence designated as the *arrival rate*. This rate represents the number of transactions per unit of time arriving at the system. After arriving, these transactions are placed in the associated queue where they wait until the server becomes empty (available) and ready for service. The selection of which transaction will receive service next is accomplished through the *queue discipline* imposed on the basic queuing system.

The discipline used to remove a transaction from the queue for service can be of several types, such as FCFS (First-Come-First-Served), LCFS (Last-Come-First-Served), SJF (Shortest-Job-First), LJF (Longest-Job-First), IS (Infinite Server), etc. A queuing discipline is nothing more than a means for choosing which transaction is to be serviced next. All queue disciplines are based on the arrival time or service time average.

A queuing model is a dynamic probabilistic model, i.e., its uses probabilities to represent the evolution, over time, of the system. It is difficult to analyze the **transient behavior** of such a model. The **transient behavior** is most readily studied via simulation. This basic queuing system is called in short a node.

For a general network system the distributions associated with each node can take on almost any form. However, to formulate a queuing-theory model as a representation of a real system, it is necessary to specify the

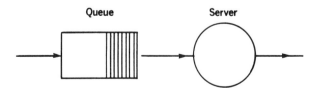

Figure 6.1 Basic Queue-Server System.

assumed form of each of the node distributions. Our choice should be based on:

1. A realistic representation of a physical queuing system.
2. A mathematically tractable system representation.

Before discussing queuing networks, we would like to study the properties of the exponential distribution and the impact that it has on queuing systems. In what follows reference is made to the basic queuing system shown in Fig. 6.1. We see that incoming transactions are stored in a queue, waiting for the server to become empty, at which time they access the server and are serviced. In the study of a system of this type we must characterize the input (transaction arrival) and the service time (service time provided by the server to the transactions). Therefore, we can state that the behavior of this basic queuing system is largely determined by two statistical properties, namely,

- **System Input** - Probability distribution of the transaction interarrival times.
- **System Service** - Probability distribution of the service times required by the transactions.

In our network model development we will assume exponential distributions for the interarrival times of transactions, as well as for the transaction service times. This distribution will fulfill the two stated conditions for the systems discussed.

We briefly review the properties of the exponential function, which was presented in Chapter 2. We know that the density function of the exponential distribution is given by

$$f(\nu) = \begin{cases} \alpha\, e^{-\alpha\, \nu} & \text{for } \nu \geqslant 0 \\ 0 & \text{for } t < 0 \end{cases} \tag{6.1}$$

and the cumulative probability of the exponential function is

$$F(t) = \int_0^t \alpha\, e^{-\alpha\, \nu}\, d\nu = -\,e^{-\alpha\, \nu}\,\Big|_0^t = 1 - e^{-\alpha\, t} \tag{6.2}$$

We can see from Equation (6.2) that $F(0) = 0$ and $F(\infty) = 1$.
The expected value is given by

$$\bar{\xi} = E[\xi] = \frac{1}{\alpha} \tag{6.3}$$

and the variance

$$var \ [\xi] \ = \ \frac{1}{\alpha^2} \tag{6.4}$$

1.1   Queue Policies

The important characteristic of a resource is the *scheduling algorithm* which decides how this resource is to be allocated to the competing transactions. There are two basic questions concerning queue scheduling:

1. Which transaction should be served now?
2. If the transaction in service is not the one we would now choose, should we allow the transaction in service to finish or should we preempt it?

The *scheduling algorithm* is also called *queue policy*, or *queue discipline*.

Some scheduling algorithms are:

- **First-Come-First-Served (FCFS)** - The choice of transaction to be served is made based on time of arrival; the transaction with the earliest arrival time is served first.
- **Last-Come-First-Served (LCFS)** - The choice of the transaction to be served is made based on time of arrival; the transaction that arrived most recently is served first.
- **Last-Come-First-Served-Preemptive-Resume (LCFSPR)** - The arriving transaction always preempts the server. If at the time of arrival, a previous transaction is being serviced, the server is preempted and the preempted transaction is placed back on the queue. This situation is illustrated in Fig. 6.2.

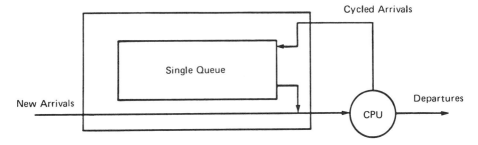

Figure 6.2   LCFSPR Scheduling Algorithm.

- **Round-Robin (RR)** - This scheduling algorithm is also called *time slicing*, which is the best known and most widely used discipline for *time-shared* computer systems. This algorithm allocates to each transaction a fixed amount of time called *quantum* or *time slice*. If the transaction's service time is completed in less than the *quantum*, it leaves; otherwise, it feeds back to the end of the queue of waiting transactions, where it waits its turn to receive another *quantum* of service, and continues in this fashion until its total service time has been satisfied. Fig. 6.3 shows this process.
- **Processor Sharing (PS)** - This particular algorithm cannot be implemented in the real system but is valuable in modeling Round-Robin scheduling. We can think of this algorithm as the limiting case of the RR algorithm (with zero overhead) as the *quantum* goes to zero. That is, all transactions are concurrently sharing the processor.
- **Shortest-Remaining-Time-First (SRTF)** - This algorithm always serves the transaction with the smallest remaining service time. If an arriving transaction has smaller service time than the transaction in service, the arriving transaction preempts the transaction in service.
- **Shortest-Latency-Time-First (SLTF)** - The operation of this algorithm is clear from its name and provides an optimal policy for drum devices.
- **Shortest-Seek-Time-First (SSTF)** - As the previous algorithm, but used for disks.

1.2 Properties of the Exponential Distribution

The exponential distribution provides several properties which are of interest for network model development [18]. We now discuss some of these.

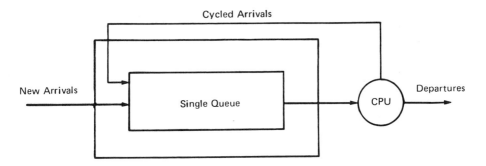

Figure 6.3 Round-Robin Scheduling Algorithm.

Property I - *The negative exponential distribution density function is a strictly decreasing function of t; as a consequence*

$$P[0 \leqslant \xi \leqslant \Delta t] > P[t \leqslant \xi \leqslant t + \Delta t] \tag{6.5}$$

for any value of t. This property can be observed by inspection; see Fig. 6.4.

Property II - *Called the Memoryless Property, establishes that the future of the exponential distribution is independent of the history of the process; therefore, this distribution remains constant in time.* This is a remarkable characteristic of the exponential distribution, which states that the probability that a transaction will finish processing in time Δt is not related to the amount of time that the transaction was in service before the start of Δt, or in other words, that the system does not remember the history of the process; thus, the system is memoryless.

Let us define the probability that a transaction which was already receiving service for t_0 finishes processing within the next t seconds.

$$P[t_0 \leqslant \xi \leqslant t + t_0] = P[\xi \leqslant t + t_0 | \xi \geqslant t_0] \, P[\xi > t_0] \tag{6.6}$$

Therefore, from Equation (6.6) we have

$$P[\xi \leqslant t + t_0 | \xi \geqslant t_0] = \frac{P[t \leqslant \xi \leqslant t + t_0]}{P[\xi > t_0]} \tag{6.7}$$

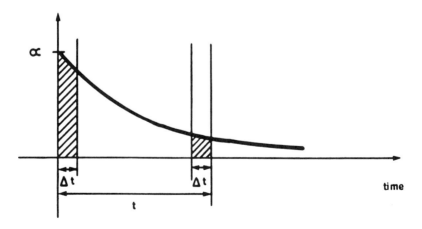

Figure 6.4   Exponential Distribution.

Since we are dealing with an exponential distribution, using Equation (6.2), we can write

$$P[\xi \leqslant t+t_0 \,|\, \xi > t] = \frac{1 - e^{-\alpha(t+t_0)} - (1 - e^{-\alpha t})}{1 - (1 - e^{-\alpha t})} \qquad (6.8)$$

from which we obtain, after some simplifications,

$$P[\xi \leqslant t+t_0 \,|\, \xi > t] = 1 - e^{-\alpha t} \qquad (6.9)$$

Property III - *The minimum of several independent exponential random variables has an exponential distribution.*
Thus,

$$\xi = min(\xi_1, \xi_2, ..., \xi_m) \qquad (6.10)$$

where ξ_i for $i=1,2,...,m$ are exponential random variables, defines a new random variable ξ, which has an exponential distribution. We know that

$$\xi_1(x) = \alpha_1 \, e^{-\alpha_1 x}$$
$$\xi_2(x) = \alpha_2 \, e^{-\alpha_2 x}$$
$$\cdots \qquad (6.11)$$
$$\xi_m(x) = \alpha_m \, e^{-\alpha_m x}$$

Therefore, using Equation (6.10), we can write

$$\xi(x) \, dx = Cont. \, [\xi_1] + Cont. \, [\xi_2] + \cdots + Cont. \, [\xi_m] \qquad (6.12)$$

where the contribution, $Cont. \, [\xi_1]$, is given by

$$Cont. \, [\xi_1] = \xi_1(x) \, dx \int_x^\infty \xi_2(v) \, dv \int_x^\infty \xi_2(v) \, dv \cdots \qquad (6.13)$$
$$\int_x^\infty \xi_m(v) \, dv$$

Thus,

$$Cont. \, [\xi_1] = \alpha_1 \, e^{-\alpha_1 x} \, e^{-\alpha_2} \cdots e^{-\alpha_m x} \qquad (6.14)$$

Adding the contributions of all the random variables, we have

$$\xi(x) = (\alpha_1 + \alpha_2 + \cdots + \alpha_m) \, e^{-(\alpha_1+\alpha_2+\cdots+\alpha_m)x} \qquad (6.15)$$

which represents an exponential distribution with the parameter given by

$$\alpha = \sum_{i=1}^m \alpha_i \qquad (6.16)$$

This property has some implications for interarrival times in queuing nodes.

Property IV- *The occurrence of n events in elapsed time τ, each exponentially distributed, is represented by a Poisson distribution.*

To prove this property, we assume that the time of occurrence of a particular event has exponential distribution with parameter α. We are interested in the number of events that occur in elapsed time τ. Let n designate the number of events. Then, the probability of having n events in time τ is given by

$$P(\xi = n; \tau) = \frac{(\alpha\tau)^n}{n!} e^{-\alpha\tau} \qquad (6.17)$$

The values of ξ have a Poisson distribution with parameter $\alpha\tau$.

The expected value of the Poisson distribution is given by

$$\bar{\xi} = E[\xi; \tau] = \alpha\tau \qquad (6.18)$$

Hence, the parameter value α is the expected value of the Poisson distribution per unit of time.

This property provides useful information about service completions. We consider two cases:

- **Case I - Single Server** - In this case n is the number of service completions achieved by a continuously busy server in elapsed time τ.
- **Case II - Multiple Server** - In this case n is the number of service completions achieved by n continuously busy servers in elapsed time τ. The parameter value is αn.

Property V - *For small time intervals, this property establishes that $k\Delta\tau$ is a good approximation to the probability that the first incident occurs in time $\Delta\tau$.*

To prove this property we expand, in a power series, the exponential given in Equation (6.18)

$$P_\xi(0,\tau) = \sum_{j=0}^{\infty} \frac{(k\Delta\tau)^j}{j!} \qquad (6.19)$$

where $\tau = \Delta\tau$. For small values of $\Delta\tau$, Equation (6.19) can be approximated by

$$P_\xi(0,\tau) \approx 1 + \Delta\tau \qquad (6.20)$$

Now, the probability that the incident will occur in $\Delta\tau$ is given by

$$P[\xi \leqslant \tau + \Delta\tau \,|\, \xi > \tau] = 1 - e^{-k\Delta\tau} \approx k\Delta\tau \qquad (6.21)$$

which proves this property.

Property VI - *Burke's Theorem* - *The departure process from a stable single-queue, with exponentially distributed service times and exponential interarrival times, is Poisson with the same parameter as the input process.* See [6,7].

To prove this theorem we make use of the *Real Convolution Theorem.*[1] *The idea behind this approach is that the output function of a system is given by the convolution of the input function and the impulse response of the system. Given two functions $f_1(t)$ and $f_2(t)$, we form the integral*

$$f(x) = \int_{-\infty}^{\infty} f_1(\xi) \, f_2(x-\xi) \, d\xi = \int_{-\infty}^{\infty} f_1(x-\xi) \, f_2(\xi) \, d\xi \qquad (6.22)$$

This integral defines a function $f(x)$ known as the convolution of $f_1(x)$ and $f_2(x)$. It is, in principle, simple enough. However, the actual computation of $f(x)$ is not trivial if $f_1(t)$ and $f_2(t)$ are not given by a simple analytic expression. Here we use the *Convolution Theorem*, which states that *if* $\phi_1(\omega)$ and $\phi_2(\omega)$ are the characteristic functions of $f_1(t)$ and $f_2(t)$, respectively, then,

$$\phi(\omega) = \phi_1(\omega) \, \phi_2(\omega) \qquad (6.23)$$

which is the characteristic function of their convolution, and is defined by

$$\phi(\omega) = \int_{-\infty}^{\infty} e^{j\omega x} f(x) \, dx \qquad (6.24)$$

The characteristic function can be used to determine the density function $f_y(y)$ of the random variable $y = g(x)$ in terms of $f_x(x)$. Let

$$\phi_y(\omega) = \int_{-\infty}^{\infty} e^{j\omega g(x)} f_x(x) \, dx \qquad (6.25)$$

But,

$$\phi_y(\omega) = \int_{-\infty}^{\infty} e^{j\omega y} f_y(y) \, dy \qquad (6.26)$$

Therefore, with a change of variable $y = g(x)$, we can write Equation (6.26) in the form

$$\phi_y(\omega) = \int_{-\infty}^{\infty} e^{j\omega y} h(y) \, dy \qquad (6.27)$$

---

[1]Papoulis, A. - *Probability, Random Variables, and Stochastic Processes*, McGraw-Hill, 1965.

where

$$f_y(y) = h(y) \qquad (6.28)$$

Therefore, generalizing this approach to the case of Poisson distributions, we can write the discrete random variables as

$$P_\xi(n) = \sum_{k=-\infty}^{\infty} P_\xi(k)\, P_\eta(n-k) \qquad (6.29)$$

which in our case becomes

$$P_\xi(n,\tau) = \sum_{k=0}^{n} \frac{(k_1\tau)^k}{k!}\, e^{-k_1\tau}\, \frac{(k_2\tau)^{n-k}}{(n-k)!}\, e^{-k_2\tau} \qquad (6.30)$$

and can be written as follows

$$P_\xi(n;\tau) = \phi(n,k_1,k_2,\tau)\, e^{-(k_1+k_2)\tau} \qquad (6.31)$$

where

$$\phi(n,k_1,k_2,\tau) = \sum_{k=0}^{n} \frac{(k_1\tau)^k}{k!}\, \frac{(k_2\tau)^{n-k}}{(n-k)!} \qquad (6.32)$$

The above equation represents a Poisson distribution, and therefore, Burke's theorem is established.

Property VII - *The sum of independent random variables with exponential distributions gives an exponential random variable with exponential distribution.*

An important application of the characteristic function is the development of a method to determine the probability density function of the sum of statistically independent variables. This method can be stated as follows: If ξ_1, ξ_2, ξ_3,....., ξ_m are mutually independent random variables for which the characteristic functions are $p_{\xi 1}$, $p_{\xi 2}$, $p_{\xi 3}$, ... , $p_{\xi m}$, then the characteristic function p_ξ for the sum is given by

$$p_\xi(t) = p_{\xi 1}\, p_{\xi 2}\, p_{\xi 3} \;\cdots\; p_{\xi m} \qquad (6.33)$$

where

$$\xi = \xi_1 + \xi_2 + \xi_3 + \cdots + \xi_m \qquad (6.34)$$

To prove this property, we consider only two functions $f_a(t)$ and $f_b(t)$, which are random variables with a Poisson distribution. The characteristic functions are given by

$$p_a(\tau) = \sum_{-\infty}^{\infty} \frac{(k_1\tau)^n}{n!} e^{-k_1\tau} e^{jn\tau} \qquad (6.35)$$

$$p_b(\tau) = \sum_{-\infty}^{\infty} \frac{e^{-k_2\tau}}{n!} e^{jn\tau} \qquad (6.36)$$

Then, the characteristic function for $f_1(t)$, defined as

$$f_1(t) = f_a(t) + f_b(t) \qquad (6.37)$$

is given by

$$p_1(\tau) = p_a(\tau) \, p_b(\tau) =$$

$$= \sum_{n=0}^{\infty} \frac{(k_1\tau)^n}{n!} e^{-k_1\tau} e^{jn\tau} \sum_{m=0}^{\infty} \frac{(k_2\tau)^m}{m!} e^{-k_2\tau} e^{jm\tau}$$

$$(6.38)$$

It is easy to see that the distribution is Poisson and that it is given by

$$P_{\xi_1}(n;\tau) = \phi(n,k_1,k_2,\tau) \, e^{-(k_1+k_2)\tau} \qquad (6.39)$$

We can generalize the above derivation for the case of m discrete random variables with Poisson distributions and show that their summation is also a Poisson distribution. Since, the exponential distribution is a Poisson distribution when $n=1$, we prove the property.

2. QUEUING NETWORKS

In the previous chapters we have discussed the birth-death process, which was a pure process, meaning that congestion was not a consideration. In this pure birth-death process we have studied several classes of structures for which we obtained the equilibrium balance condition. The motivation of this presentation was to learn about the behavior of these chains of state changes, which we define as *Markov Chains*. The behavior of real systems is characterized by the presence of several congestion points. These congestion points are a result of sharing several types of resources (servers). At these congestion points a queue is formed where transactions are waiting to receive service. The collection of these congestion points and the interconnection between them constitute a **Queuing Network** (QN). A good presentation of these concepts, as well as the introduction to the different types of computer models, is available in a book by Ajmone Marsan and others [1]. This presentation is concise and easy to understand.

2.1   Feedforward Queuing Networks

For this type of networks, the output of a processing station will be the input to one or more processing stations. No looping is allowed, and therefore, the flow propagation is always forward. We represent these networks with the characteristics indicated in Fig. 6.5.

We will consider the following topics, which fall under the classification of QN,

- Jackson Queuing Networks
- Gordon and Newell Queuing Networks
- BCMP Queuing Networks
- Buzen's Queuing Networks
- Mean Value Analysis Approach
- Latest Developments in Analyzing Queuing Networks

2.1.1   Jackson Queuing Networks

We are now interested in extending the discussion to **multidimensional birth-death processes**. For this purpose, a convenient and practical model for a processing network is a network of queues, in which we make the assumption that a queue precedes each server (processing station). The output of a queue is then fed to another queue, after receiving service. These particular networks can be modeled and analyzed by means of mul-

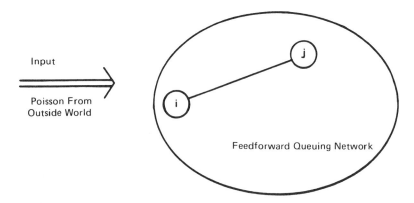

Figure 6.5   Feedforward Queuing Networks.

tidimensional birth-death processes. We call these networks **Jackson Networks**, named after J. R. Jackson who discovered the product form solution [20,21].

We are going to present and discuss the following:

- Queues in Tandem Networks
- Queues in Parallel Networks
- Queues in Acyclic Networks
- Queues in Feedback Networks
- Local Computer Networks

Queues in Tandem Networks - The theory of Jackson networks is based on multidimensional birth-death processes, which is a straight forward extension of the one-dimensional birth-death process studied. Consider the two queues in tandem shown in Fig. 6.6. The transactions arrive at Queue 1, Poisson distributed, with rate λ, and are queued in this queue awaiting processing. When the server for this queue finishes service on a given transaction, this serviced transaction moves to Queue 2, and a new transaction from Queue 1 moves to the server to start service.

We define

$$Q_1(\tau) = \textit{number of transactions in Queue 1}$$

$$Q_2(\tau) = \textit{number of transactions in Queue 2}$$

$$\mu_1 = \textit{service rate at server 1}$$

$$\mu_2 = \textit{service rate at server 2}$$

$$\lambda = \textit{transaction arrival rate}$$

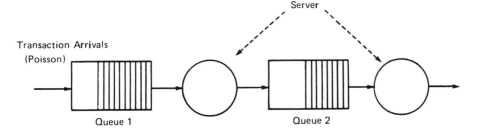

Figure 6.6 Two Queues in Tandem.

We are interested in the joint probability of having a queue length of η_1 in Queue 1 and a queue length of η_2 in Queue 2, which can be expressed as

$$P(\eta_1,\eta_2,\tau) = P[Q_1(\tau)=\eta_1,Q_2(\tau)=\eta_2] \tag{6.40}$$

Since we are interested in obtaining the expression of Equation (6.40), we need to first obtain the differential equation of the process and the probabilities involved in an incremental time, such as $\Delta\tau$. For an interval of time $\Delta\tau$, we consider the following four cases:

- $P(0,0;\tau+\Delta\tau)$ - Probability of having zero queue length in Queue 1 and Queue 2, in the interval $\Delta\tau$.
- $P(0,\eta_2;\tau+\Delta\tau)$ - Probability of having zero queue length in Queue 1 and η_2 queue length in Queue 2, in the interval $\Delta\tau$.
- $P(\eta_1,0;\tau+\Delta\tau)$ - Probability of having η_1 queue length in Queue 1 and zero queue length in Queue 2, in the interval $\Delta\tau$.
- $P(\eta_1,\eta_2;\tau+\Delta\tau)$ - Probability of having η_1 queue length in Queue 1 and η_2 queue length in Queue 2, in the interval of time $\Delta\tau$.

The probability of having zero queue lengths in the time interval $\Delta\tau$, from τ to $\tau+\Delta\tau$, is composed of two probabilities

$$P(0,0;\tau+\Delta\tau) = P(0,1;\tau)\ \mu_2\Delta\tau\ +\ (0,0;\tau)\ (1-\lambda\Delta\tau) \tag{6.41}$$

These are the only two cases that can occur in the time interval $\Delta\tau$, which are 1. Having a queue length of 1 at Queue 1. During $\Delta\tau$, the transaction in the queue moves to the server, since this was empty. 2. Having zero queue lengths in all queues.

Similarly, for the other equations we can write

$$P(0,0;\tau+\Delta\tau) = P(0,1;\tau)\ \mu_2\Delta\tau\ +\ (1-\lambda\Delta\tau)\ P(0,0;\tau)$$

$$P(0,\eta_2;\tau+\Delta\tau) = P(1,\eta_2-1;\tau)\ \mu_1\Delta\tau$$
$$+\ P(0,\eta_2+1;\tau)\ \mu_2\Delta\tau\ +\ [1-(\lambda+\mu_2)\Delta\tau]\ P(0,\eta_2;\tau)$$

$$P(\eta_1,0;\tau+\Delta\tau) = P(\eta_1-1,0;\tau)\ \mu_1\Delta\tau$$
$$+\ P(\eta_1,1;\tau)\ \mu_2\Delta\tau\ +\ [1-(\lambda+\mu_1)\Delta\tau]\ P(\eta_1,0;\tau)$$

$$P(\eta_1,\eta_2;\tau+\Delta\tau) = P(\eta_1+1,\eta_2-1;\tau)\ \mu_1\Delta\tau$$
$$+\ P(\eta_1-1,\eta_2;\tau)\ \lambda\Delta\tau\ \ P(\eta_1,\eta_2+1;\tau)\ \mu_2\Delta\tau$$
$$+\ [1-(\lambda+\mu_1+\mu_2)\Delta\tau]\ P(\eta_1,\eta_2;\tau)$$

Since the derivative of a function f(t) is given by

$$f'(t) = \lim_{\Delta t \to 0} \frac{\Delta y(t)}{\Delta t} = \lim_{\Delta t \to 0} \frac{f(t+\Delta t) - f(t)}{\Delta t} \tag{6.42}$$

The first equation can be written as follows

$$P(0,0;\tau+\Delta\tau) - P(0,0;\tau) = P(0,1;\tau)\,\mu_2\Delta\tau - \lambda\Delta\tau\,P(0,0;\tau) \tag{6.43}$$

or,

$$\frac{P(0,0;\tau+\Delta\tau) - P(0,0;\tau)}{\Delta\tau} = P(0,0;\tau)\mu_2 - \lambda P(0,0;\tau) \tag{6.44}$$

Taking the limit on both sides yields

$$\lim_{\Delta\tau \to 0} \left[\frac{P(0,0;\tau+\Delta\tau) - P(0,0;\tau)}{\Delta\tau} \right] \tag{6.45}$$

$$= \lim_{\Delta\tau \to 0} \left[P(0,1;\tau)\mu_2 - \lambda P(0,0;\tau) \right]$$

The left-hand side is the derivative of $P(0,0;\tau)$, and the right-hand side, which is not a function of $\Delta\tau$, remains the same in the limiting process; thus,

$$\frac{dP(0,0;\tau)}{d\tau} = \mu_2 P(0,1;\tau) - \lambda P(0,0;\tau) \tag{6.46}$$

With similar reasoning, we get the system of differential equations:

$$\frac{dP(0,0;\tau)}{d\tau} = P(0,1;\tau)\mu_2 - P(0,0;\tau)\lambda$$

$$\frac{dP(0,\eta_2;\tau)}{d\tau} = P(1,\eta_2-1;\tau)\,\mu_1 + P(0,\eta_2+1;\tau)\,\mu_2$$

$$- P(0,\eta_2;\tau)(\lambda+\mu_2)$$

$$\frac{dP(\eta_1,0;\tau)}{d\tau} = P(\eta_1-1,0;\tau)\lambda + P(\eta_1,1;\tau)\,\mu_2$$

$$- P(\eta_1,0;\tau)(\lambda+\mu_1)$$

$$\frac{dP(\eta_1,\eta_2;\tau)}{d\tau} = P(\eta_1+1,\eta_2-1;\tau)\,\mu_1 + P(\eta_1-1,\eta_2;\tau)\lambda$$

$$+ P(\eta_1,\eta_2+1;\tau)\,\mu_2 - P(\eta_1,\eta_2;\tau)(\lambda+\mu_1+\mu_2)$$

By definition, an *equilibrium state probability distribution* satisfies

$$\lim_{\tau \to \infty} P(\eta_1,\eta_2;\tau) = P(\eta_1,\eta_2) \qquad (6.47)$$

The equilibrium state probability distribution (steady state) provides a description of the long-run-average behavior of the state of the system. Under this condition all derivatives of $P(\eta_1,\eta_2;\tau)$, with respect to time, are null. Therefore, we can write

$$P(0,1;\tau)\,\mu_2 - P(0,0;\tau)\lambda = 0$$

$$P(1,\eta_2-1;\tau)\,\mu_1 + P(0,\eta_2+1;\tau)\,\mu_2 - P(0,\eta_2;\tau)(\lambda+\mu_2) = 0$$

$$P(\eta_1-1,0;\tau)\lambda + P(\eta_1,1;\tau)\,\mu_2 - P(\eta_1,0;\tau)(\lambda+\mu_1) = 0$$

$$P(\eta_1+1,\eta_2-1;\tau)\,\mu_1 + P(\eta_1-1,\eta_2;\tau)\lambda$$
$$+ P(\eta_1,\eta_2+1;\tau)\,\mu_2$$
$$- P(\eta_1,\eta_2;\tau)(\lambda+\mu_1+\mu_2) = 0$$

If $\lambda < \mu_1$ and $\lambda < \mu_2$, a steady solution $P(\eta_1,\eta_2)$ exists. In the equilibrium case it is clear that the flow must be conserved, in the sense that for each state the input flow must be equal to the output flow. If we delete the variable τ from the equations, we have

$$P(0,1)\mu_2 - P(0,0)\lambda = 0$$

$$P(1,\eta_2-1)\,\mu_1 + P(0,\eta_2+1)\,\mu_2 - P(0,\eta_2)(\lambda+\mu_2) = 0$$

$$P(\eta_1-1,0)\lambda + P(\eta_1,1)\,\mu_2 - P(\eta_1,0)(\lambda+\mu_1) = 0$$

$$P(\eta_1+1,\eta_2-1)\,\mu_1 + P(\eta_1-1,\eta_2)\lambda + P(\eta_1,\eta_2+1)\,\mu_2$$
$$- P(\eta_1,\eta_2)(\lambda+\mu_1+\mu_2) = 0$$

We now analyze the last equation, and for convenience the equation is repeated below

$$P(\eta_1+1,\eta_2-1)\,\mu_1 + P(\eta_1-1,\eta_2)\lambda + P(\eta_1,\eta_2+1)\,\mu_2$$
$$- P(\eta_1,\eta_2)(\lambda+\mu_1+\mu_2) = 0 \qquad (6.48)$$

Looking at Equation (6.48), we can observe five states

$$S(\eta_1-1,\eta_2); \ S(\eta_1,\eta_2); \ S(\eta_1+1,\eta_2); \ S(\eta_1,\eta_2+1);$$
$$\text{and } S(\eta_1-1,\eta_2+1) \qquad (6.49)$$

We also observe that any new arrival will move the order of the state to a new one, which is accomplished by increasing the number of transactions in Queue 1. This is the only mechanism in this system that can upgrade the state. This situation is shown in Fig. 6.7.

Since for any system in equilibrium, the flow rate that comes into a given state must be equal to the flow rate that goes out of the state, to obtain the equilibrium equation we must find all the flows going into a generic state, as well as all the flows going out of this state. Thus, observing Fig. 6.7, we can write

$$Flow\ rate\ in = P(\eta_1-1,\eta_2)\ \lambda + P(\eta_1,\eta_2+1)\ \mu_2$$
$$+ P(\eta_1+1,\eta_2)\ \mu_1 \tag{6.50}$$
$$Flow\ rate\ out = P(\eta_1,\eta_2)\ \lambda + P(\eta_1,\eta_2)(\lambda+\mu_1+\mu_2) \tag{6.51}$$

Setting

$$Flow\ rate\ in = Flow\ Rate\ out \tag{6.52}$$

we obtain Equation (6.48). The solution of these equilibrium equations has the form

$$P(\eta_1,\eta_2) = P_1(\eta_1)\ P_2(\eta_2) \tag{6.53}$$

The normalization condition is

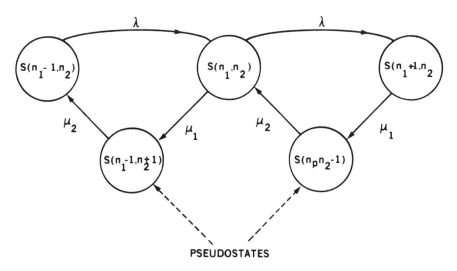

Figure 6.7   State Diagram.

$$\sum_{\eta_1=0}^{\infty} \sum_{\eta_2=0}^{\infty} P(\eta_1,\eta_2) = 1.0 \tag{6.54}$$

Generalizing Equation (6.53) for N queues in tandem, we have

$$P(\eta_1,\eta_2,\ldots,\eta_N) = P_1(\eta_1) \, P_2(\eta_2) \, \ldots \, P_N(\eta_N) \tag{6.55}$$

Vital to the derivation of the product form solution, which is the one provided by Equation (6.55), was the application of Burke's theorem, made possible by the fact that no cycles were present in the network. Burke's theorem fails to hold as soon as feedback exists between the output and the input of a server. The pioneering work of Jackson shows that the product form solution holds for a more general class of networks in which a transaction is allowed to receive service more than one time in a given processor.

Jackson's Theorem - *If the system of equations balancing the flows of transactions entering and leaving a processor has a solution, and if this solution is such that the stability conditions are satisfied, then the product form solution holds.*

The model is analyzed by recognizing that the true behavior of the state of the network is a Markov chain and that the equations for the steady-state solution of the model can be written by balancing the probability of entering a state with the probability of exiting from that same state.

We use the following equation to define the state of the network,

$$S = (n_1,n_2,\ldots,n_N)$$

where the values n_i indicates the number of transactions at processor i.
The following are the only transitions allowed in the network:

$n_{0j} \rightarrow S$ a transaction arrives from the outside world to processor j
$n_{i0} \rightarrow S$ a transaction departs from processor i and leaves the network
$n_{ij} \rightarrow S$ a transaction departs processor i and joins processor j

Queues in Parallel Networks - It can also be shown that these results hold when there is more than one server at the facility, representing the case of parallel networks.
Queues in Acyclic Networks - Consider the network shown in Fig. 6.8. The characteristics of this network are:

- Messages arrive at each of the four nodes with independent rates and with Poisson distribution.
- At each of the nodes the service time is independent of the queue length

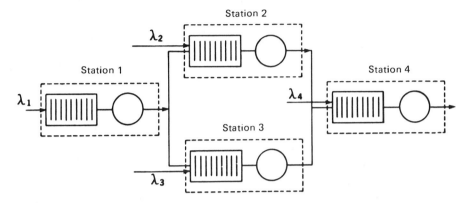

Figure 6.8   Acyclic Network.

as well as the other node's service times. This service time is exponentially distributed.

- The output of service station 1 goes to the queue of the processing station 2 with probability α. Similarly, the output goes to station 3 with a probability $1 - \alpha$.

Under these conditions, we can show that the sum of Poisson processes is Poisson and that the random bifurcation (α and $1 - \alpha$) of Poisson processes yields a Poisson distribution. Furthermore, from Burke's theorem the output of an M/M/m queue is Poisson. From all these considerations the flows within the network are Poisson. Therefore, we expect the product form to hold for networks with individual queues behaving as M/M/m.

Queues in Feedback Networks - Consider the network shown in Fig.

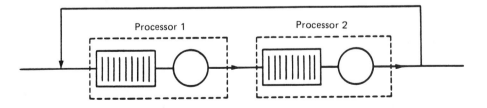

Figure 6.9   Feedback Network.

6.9. The situation is not as straightforward when there is a feedback path. The presence of a feedback path destroys the Poisson character of the flow within the network; see [39]. It can be shown that the flow on links between queues within a feedback loop is not Poisson. Nevertheless, as we shall see, the product form still holds.

2.1.2 Gordon and Newell Queuing Networks

A special case of Jackson networks are those in which no interaction with the outside world is allowed; see [20]. Networks of this type can be used to model systems in which new transactions enter only upon the departure of another transaction with the same characteristics. These are called **Closed Networks**. Fig. 6.10 (a) and (b) shows, respectively, **open networks** and **closed networks**.

Open Networks - Networks in which the output of the network is not related to the input.

Closed Networks - The output to the network is closely related to the input. So closely related, that for each transaction which finished processing in the network, a new one is created in the input with the same characteristics. This is the type of network that we will use and discuss.

2.1.3 BCMP Queuing Networks

Several extensions of the work of Jackson and of Gordon and Newell have been proposed, all leading to QN with *product form solution*. These extensions known as BCMP Queuing Networks (Baskett, Chandy, Muntz and Palacios Queuing Networks) are concerned with the possibility of grouping the transactions of the network into classes with different characteristics. For a complete discussion of this subject, see [3].

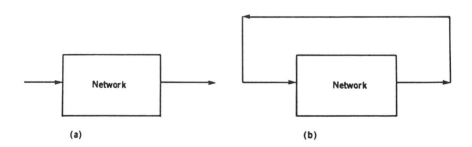

(a)   (b)

Figure 6.10   Types of Networks.

Local Balance Queuing Networks - We now apply the material covered in the previous chapters, particularly the one on Markovian processes and associated balance equations, as applied to local computer networks. For the Jackson networks we obtained a **product form**, and now we are interested to see under what conditions a product form solution exists. To have a product form solution, a queuing network must be composed of service stations (processors), each one satisfying the **local balance property**.

For a better understanding of the state diagram development methodology consider the local computer network shown in Fig. 6.11. This network is composed of a CPU and two I/O units. Only one class of transactions is considered. The service provided by this network is indicated by the rate in each processor, denoted τ_i, $i=1,2,3,4$. The transfer probabilities between the CPU and the I/O units are p_{0i}, with $i=1,2$. Also, the output of the I/O1 device is split between two different paths; one goes to the CPU queue, while the other goes to the I/O2 queue, with probabilities p_{21} and p_{23}, respectively. These probabilities satisfy the following conditions:

$$p_{12} + p_{13} = 1.0$$
$$p_{21} + p_{23} = 1.0$$

The state of the network is designated as

$$S = (n_1, n_2, n_3, n_4) \qquad (6.56)$$

If we have only four transactions in the system, we can write the state diagram as in Fig. 6.12.

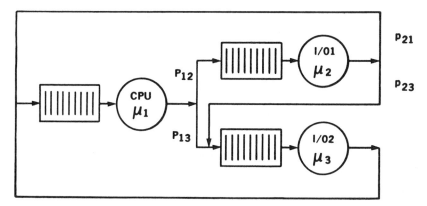

Figure 6.11   Local Computer Network.

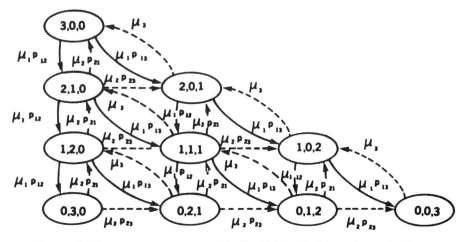

Figure 6.12 State Diagram (copyright © 1986 by The Massachusetts Institute of Technology reprinted by permission of MIT Press).[2]

To this end we would like to define the birth rate, which is given by

$$\mu_i = \frac{1}{\tau_i} \qquad (6.57)$$

In writing Equation (6.57) we must remember that the death rate is the inverse of the service time. To have a clear diagram of the state transitions, some of the transition paths are ignored in Fig. 6.12. The transition paths that are not shown are:

$$S(1,1,1) \rightarrow S(2,1,0) \qquad \mu_3$$
$$S(2,1,0) \rightarrow S(1,1,1) \qquad \mu_1\,p_{13}$$
$$S(0,2,1) \rightarrow S(1,2,0) \qquad \mu_3$$
$$S(1,2,0) \rightarrow S(0,2,1) \qquad \mu_1\,p_{13}$$
$$S(0,1,2) \rightarrow S(1,1,1) \qquad \mu_3$$
$$S(1,1,1) \rightarrow S(0,1,2) \qquad \mu_1\,p_{13}$$

We first discuss how the state diagram was developed. We assume that the network is a closed network, holding three transactions. For this particular network and this population we write all the possible states. When

[2] Reprinted from the book by M. Ajmon Marsan, G. Balbo, and G. Conte - *Performance Models of Multiprocessor Systems*, MIT Press, 1986.

this is done, we locate each state in such a way that all the possible state transitions can be shown in a clear manner, as in Fig. 6.12. The state with one transaction in each device is $S(1,1,1)$. This state is identified in Fig. 6.12 with the double dashed rectangle. With reference to this state we can write the *global balance equation*

$$P(1,1,1) \ (\mu_1 + \mu_2 + \mu_3) \ =$$

$$= P(2,1,0) \ \mu_1 p_{13} \ + \ P(1,2,0) \ \mu_2 p_{23}$$

$$+ \ P(2,0,1) \ \mu_1 p_{12}$$

$$+ \ P(0,2,1) \ \mu_2 p_{21}$$

$$+ \ P(0,1,2) \ \mu_3$$

A sufficient condition for the state probability distribution to satisfy this global balance equation is that the following flows be individually balanced

$$P(1,1,1) \ \mu_3 = P(2,1,0) \ \mu_1 p_{13} \ + \ P(1,2,0) \ \mu_2 p_{21}$$
$$P(1,1,1) \ \mu_1 = P(0,2,1) \ \mu_2 p_{21} \ + \ P(0,1,2) \ \mu_3$$
$$P(1,1,1) \ \mu_2 = P(2,0,1) \ \mu_1 p_{12}$$

Splitting the global balance equation into *local balance equations* amounts to the construction of a larger system of linear equations that is usually easier to solve. Moreover, it is not difficult to show that if such a system is compatible, so that a solution exists, the solution has a product form.

2.1.4   Buzen's Queuing Networks

Following Buzen's development in [8], we will show that all the network performance indices can be derived from the G coefficients of the generating function of the network. Before discussing the generating function of the processing network, let us state clearly the assumptions we need in the processing network to ease the mathematical manipulations. For more information on this subject, see [8,9,15,16,40]. Some software packages have been developed using this technique [4,17]. The assumptions are:

- **Exponential Distribution** - The holding time at each processing station is an *exponentially distributed random variable*, so that at any instant the probability of the transaction finishing within the next Δt time interval is just $\Delta t / \tau$. Thus, this probability is independent of how long the transaction has been processing.

• **Work Conservative** - The number of transactions arriving at a processing station is (approximately) the same as the number departing.

• **Homogeneity** - On-line processing functions of the stations are the same as the off-line processing functions.

• **Network Type** - The processing network is assumed to be a *closed-network*.

• **Finite Period** - All quantities are based on a finite observation period.

If we concentrate on only one performance index, such as throughput, we can make use of Linear Programming (LP) for the mathematical model description. If, on the other hand, we would like to calculate several indices, such as throughput, utilization, queue lengths, etc., at capacity level, the use of queuing theory is almost mandatory.

Transfer Probabilities - We can represent a queuing system as a finite number of processing stations at which a given transaction receives service. Each of these processing stations is connected to a number of other stations, according to the movement of transactions. Let us call these connectivity paths the processing network. A schematic of a portion of the processing network is represented in Fig. 6.13. Observing this figure, we see that when the processing of a given transaction is finished at processing station i, it seeks processing at stations j, k or l, with probabilities p_{ij}, p_{ik} and p_{il}, respectively, where p_{ij} is the probability that the transaction moves to station j after being serviced at station i. Let N denote the number of processing stations in the manufacturing plant.

We define the *relative frequency*[3] for service station i, ρ_i, as the mean arrival rate of transactions to that service station. It is important to note that there is no unique solution to the values of ρ_i and that the ρ_i's can only be determined to within a multiplicative constant. It is also important to note that when we use the concept of maximum utilization (discussed later), all the values of ρ_i are identical.

We now calculate the relative frequencies defined as solutions to:

$$\rho_j = \sum_{i=1}^{N} \rho_i \, p_{ij} \tag{6.58}$$

where

---

[3] Kleinrock, L. - Computer Network Research - Technical Report, School of Engineering and Applied Science, University of California, Los Angeles, California, December 31, 1975.

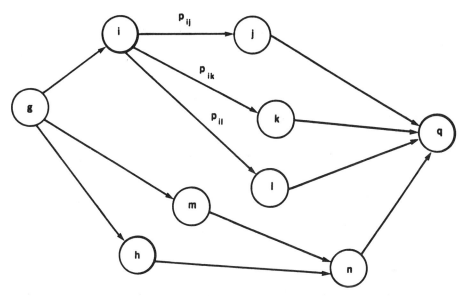

Figure 6.13   Generic Network Processing Configuration.

$$\rho_i = \text{relative frequency at processing station } i$$

$$p_{ij} = \text{probability that when a transaction finishes at } i,$$
$$\text{it seeks service at } j$$

It is interesting to observe that the system of equations defined by Equation (6.58) is homogeneous and hence admits an infinite number of solutions. These solutions differ by a multiplication constant, and any one of them is equally suited for computing the product form solution of the network in which the influence of the multiplication constant is canceled.

Maximum Utilization Concept - To assure the maximum utilization of the processors, the transfer probabilities must be such that the utilization of devices j, k, and l are the same. The average service time of the transactions at these devices is τ_j, τ_k, and τ_l. Therefore, to assure maximum utilization

$$n_j \tau_j = n_k \tau_k = n_l \tau_l \tag{6.59}$$

where n_j, n_k, and n_l are the number of transactions processed in devices j, k, and l during time $n_j \tau_j$. These are real numbers. Fixing one of them, we can calculate the others.

$$n_k = \frac{n_j \tau_j}{\tau_k} \qquad (6.60)$$

$$n_l = \frac{n_j \tau_j}{\tau_l} \qquad (6.61)$$

Then the probabilities for maximum utilization are given by

$$p_{ij} = \frac{n_j}{n_j + n_k + n_l} \quad ; \quad p_{ik} = \frac{n_k}{n_j + n_k + n_l} \quad ; \quad p_{il} = \frac{n_l}{n_j + n_k + n_l} \quad (6.62)$$

Let us present an example to illustrate the calculation of these probabilities.

Example 6.1 - Let us consider the computer network shown in Fig. 6.14. We will assume that $p_{00} = 0$. We are interested in obtaining the transfer probabilities assuming that the service times are:

$$\tau_0 = 3.20$$
$$\tau_1 = 3.85$$
$$\tau_2 = 4.21$$

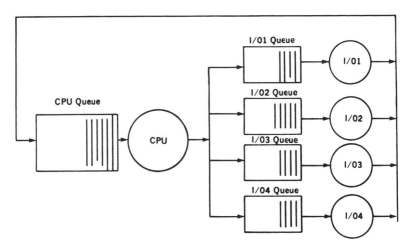

Figure 6.14   Local Computer Network.

$$\tau_3 = 3.26$$
$$\tau_4 = 3.72$$

In our case we have

$$n_1\tau_1 = n_2\tau_2 = n_3\tau_3 = n_4\tau_4$$

Assuming $n_1 = 1$, we obtain

$$n_2 = \frac{n_1\tau_1}{\tau_2} = \frac{3.85}{4.21} = 0.9144$$

$$n_3 = \frac{n_1\tau_1}{\tau_3} = \frac{3.85}{3.26} = 1.1809$$

$$n_4 = \frac{n_1\tau_1}{\tau_4} = \frac{3.85}{3.72} = 1.0349$$

The total average number of processing transactions is

$$\Delta = n_1 + n_2 + n_3 + n_4$$
$$= 1 + 0.9144 + 1.1809 + 1.0349$$
$$= 4.1302$$

The transfer probabilities are

$$p_{01} = \frac{n_1}{\Delta} = \frac{1}{4.1302} = 0.2421$$

$$p_{02} = \frac{n_2}{\Delta} = \frac{0.9144}{4.1302} = 0.2213$$

$$p_{03} = \frac{n_3}{\Delta} = \frac{1.1809}{4.1302} = 0.2859$$

$$p_{04} = \frac{n_4}{\Delta} = \frac{1.0349}{4.1302} = 0.2505$$

Therefore, to obtain the full utilization of the system the transfer probabilities between the CPU and the four I/O devices must be the ones indicated above.

The relative frequencies ρ_i of these probabilities are given as the solution to the following system of equations:

$$\sum_{i=1}^{4} \rho_i p_{ji} = \rho_j \qquad j=1,2,3,4$$

Therefore, we have

$$\rho_1 = 0.2421 \, \rho_1 + 0.2213 \, \rho_2 + 0.2859 \, \rho_3 + 0.2505 \, \rho_4$$
$$\rho_2 = 0.2421 \, \rho_1 + 0.2213 \, \rho_2 + 0.2859 \, \rho_3 + 0.2505 \, \rho_4$$
$$\rho_3 = 0.2421 \, \rho_1 + 0.2213 \, \rho_2 + 0.2859 \, \rho_3 + 0.2505 \, \rho_4$$
$$\rho_4 = 0.2421 \, \rho_1 + 0.2213 \, \rho_2 + 0.2859 \, \rho_3 + 0.2505 \, \rho_4$$

which can be written as follows

$$-0.7579 \, \rho_1 + 0.2213 \, \rho_2 + 0.2859 \, \rho_3 + 0.2505 \, \rho_4 = 0$$
$$0.2421 \, \rho_1 - 0.7787 \, \rho_2 + 0.2859 \, \rho_3 + 0.2505 \, \rho_4 = 0$$
$$0.2421 \, \rho_1 + 0.2213 \, \rho_2 - 0.7141 \, \rho_3 + 0.2505 \, \rho_4 = 0$$
$$0.2421 \, \rho_1 + 0.2213 \, \rho_2 + 0.2859 \, \rho_3 - 0.7495 \, \rho_4 = 0$$

From inspection of the above system of equations, we see that the solution is obtained when all the ρ's are the same. Thus,

$$\rho_1 = \rho_2 = \rho_3 = \rho_4$$

Therefore, we take the values of $\rho_i = 1.0$ for $i = 1,2,3,4$. Thus, we see that the concept of maximum capacity makes all relative frequencies equal, which agrees with the concept of maximum utilization.

For a better understanding of the methodology for model development and, in particular, in the evaluation of system performance, we consider the following cases:

- Case I - One class of transaction
- Case II - Two classes of transactions
- Case III - Multiple classes of transactions

3. CASE I: ONE CLASS OF TRANSACTIONS

In this case we consider a closed network holding n transactions of the same class, each receiving service at the different network processors. The state of the network at any time is denoted by

$$S = (n_1, n_2, \cdots, n_N) \qquad (6.63)$$

where the values of n_i are given by

$n_i = 0$ Processor i idle
$n_i = 1$ Transaction processing at processor i
$n_i = 2$ One transaction processing and one transaction waiting
            ...
$n_i = k$ One transaction processing and $k-1$ transactions waiting, $k \geqslant 1$

Since we assume a closed network, the total number of transactions remains constant and is given by

$$\sum_{i=1}^{N} n_i = n \qquad (6.64)$$

From Fig. 6.11, we see that the probability associated with a transaction moving from the ith node to the jth node is given by

$$\frac{\Delta t}{\tau_i} p_{ij} P_r(S') \qquad (6.65)$$

where S' represents the new state after the transaction has moved from the ith node to the jth node. This state is given by

$$S' = (n_1, n_2, n_3, \cdots, n_i - 1,, \cdots, n_j + 1, \cdots, n_N) \qquad (6.66)$$

In the above equation, we see that the number of transactions at the ith processor has been decreased by one, and the number of transactions at the jth processor has been increased by one. This statement represents the move of a transaction from the ith processor to the jth processor. Generalizing, we see that the probability of leaving the ith node, for any i, and going to a node connected to node i is given by

$$\sum_{j=1}^{N} \frac{\Delta t}{\tau_i} p_{ij} P_r(S') \qquad (6.67)$$

Therefore, the basic equilibrium relation in the network is given by

$$\sum_{i/n_i>0} \frac{\Delta t}{\tau_i} P_r(S) = \sum_{i/n_i>0}^{N} \sum_{j=1}^{N} \frac{\Delta t}{\tau_i} p_{ij} P_r(S') \qquad (6.68)$$

The basic transition relation for equilibrium in the network can be stated as follows: *For any state S the probability $P_r(S)$, of being in that state times the probability of a transition from that state, has to be equal to the sum over all states S' of $P_r(S')$ times the probability of a transition from S' to S.*

Therefore, we can write Equation (6.68) as follows

$$\sum_{j/n_j>0}^{N} \frac{1}{\tau_j} P_r(n_1, n_2, \cdots, n_N) =$$

$$\sum_{j/n_j>0}^{N} \sum_{i=1}^{N} \frac{p_{ij}}{\tau_i} P_r(n_1, n_2, \cdots, n_{n+1}, \cdots, n_{j-1}, \cdots, n_N) \qquad (6.69)$$

Our problem is to find the solution to Equation (6.69). A solution is given by

$$P_r(n_1, n_2, \cdots, n_N) = \frac{1}{C} \prod_{i=1}^{N} x_i^{n_i} \qquad (6.70)$$

where $x_i = \rho_i \, \tau_i$, and the normalizing constant C is determined by the requirement that the probabilities summed over all states be unity.

$$C = \sum_{i=1}^{N} \prod_{i=1}^{N} x_i^{n_i} \qquad (6.71)$$

Now, if we look at the polynomial

$$g(t) = (1 + x_1 t + x_1^2 t^2 + \ldots)(1 + x_2 t + x_2^2 t^2 + \cdots)$$
$$\cdots (1 + x_N t + x_N^2 t^2 + \ldots.) \qquad (6.72)$$

clearly, the coefficients of t^n in $g(t)$ are just the normalized constant C, since that coefficient is the result of a sum of terms of the form $x_1^{n_1}, x_2^{n_2}, \ldots, x_N^{n_N}$ with $\sum n_i = n$.

3.1 Generating Function of the Network

We will call $g(t)$ the *generating function of the network*; its factors will be called the generating functions of the individual processors. Let

$$g(t) = 1 + G(1)t + G(2)t^2 + G(3)t^3 + \cdots \qquad (6.73)$$

We examine just how the product $g(t)$ is constructed. Let the factors of $g(t)$ be denoted by $x_i(t)$.

$$x_1(t) = 1 + x_1 t + x_1^2 t^2 + x_1^3 t^3 + \cdots$$
$$x_2(t) = 1 + x_2 t + x_2^2 t^2 + x_2^3 t^3 + \cdots$$
$$\cdots$$
$$x_N(t) = 1 + x_N t + x_N^2 t^2 + x_N^{t3} + \cdots$$

The first term factor of the $x_1(t)$ is 1. This term, when multiplied by the other factors, yields terms of the form $x_1^0 \, x_2^{n_2} \, x_3^{n_3} \ldots x_N^{n_N}$. If this term is left out of $x_1(t)$, and the polynomial product recalculated, we see that the coefficient of t^n is just the sum of all terms for which $n_1 \geqslant 1$. The coefficient of t^n in the recomputed product divided by the coefficient of t^n in $g(t)$ is just the probability that $n_1 \geqslant 1$, i.e., it is the utilization of processor 1, or the fraction of time processor 1 is busy.

$$h_1(t) = g(t) \frac{x_1(t) - 1}{x_1(t)}$$

$$= g(t) \left[1 - \frac{1}{x_1(t)} \right] \tag{6.74}$$

Let us call the function $h_1(t)$ the generating function of the utilization. But,

$$x_1(t) = 1 + x_1 t + x_1^2 t^2 + \cdots = \frac{1}{1 - x_1 t} \tag{6.75}$$

Thus,

$$h_1(t) = x_1 t g(t) \tag{6.76}$$

3.2 Algorithm for Calculating the G_i

The generating function of the network is given by

$$g(t) = \prod_{j=1}^{N} (1 + x_j t + x_j^2 t^2 + \cdots) \tag{6.77}$$

Let us designate

$$g_1(t) = x_1(t) = 1 + x_1 t + x_1^2 t^2 + \cdots$$

$$g_2(t) = g_1(t)(1 + x_2 t + x_2^2 t^2 + \cdots) = g_1(t) x_2(t)$$

And in general we can write

$$g_i(t) = g_{i-1} x_i(t) \tag{6.78}$$

But since

$$x_i(t) = \frac{1}{1 - x_i t} \tag{6.79}$$

we can write

$$g_i(t) = g_{i-1}(t) + x_i t g_i(t) \tag{6.80}$$

Let $G_i(j)$ be the coefficient of t^j in $g_i(t)$. Then

$$G_i(j) = G_{i-1}(j) + x_i G_i(j-1) \tag{6.81}$$

which is an efficient and simple algorithm for the calculation of $G_i(j)$. Two important consequences can be drawn for Equation (6.81):

- All the $G_i(0)$ are equal to one, or

$$G_i(0) = 1 \qquad (i = 1, 2, \ldots, N)$$

- $G_1(j)$ are given by

$$G_1(j) = x_1 \, G_1(j-1)$$

3.2.1   Algorithm Implementation

The implementation of the iterative Equation (6.81), which is simple, is as follows:

1. **Matrix Definition** - Define a matrix (G-matrix) of size $(n+1)N$, where n is the number of transactions and N is the number of processors in the network.
2. **First Row** - All the elements in the first row of the G-matrix are 1, since $G_i(0) = 1$, for $i = 1, 2, ..., N$.
3. **First Column** - All the elements in the first column of the G-matrix are calculated by the formula $x_1{}^j$, for $j = 1, 2, ..., n+1$, since for this column Equation (6.81) becomes

$$G_1(j) = x_1 \, G_1(j-1) = x_1{}^j$$

4. G-matrix - All the elements from row 2 to N, and column 2 to $n+1$ are determined by the rule: *each element in the G-matrix, from*

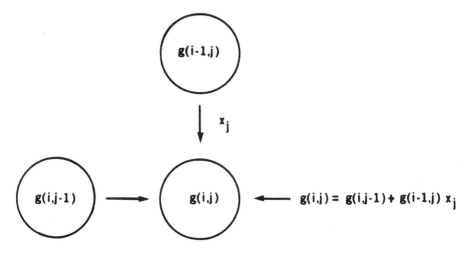

Figure 6.15   Illustration of the Calculation Process.

Table 6.1 Coefficients of the Generating Function

Powers of t	Processing Stations 1				
0	$G_1(0)$	$G_2(0)$	$G_3(0)$	...	$G_s(0)$
1	$G_1(1)$	$G_2(1)$	$G_3(1)$	...	$G_s(1)$
2	$G_1(2)$	$G_2(2)$	$G_3(2)$	...	$G_s(2)$
3	$G_1(3)$	$G_2(3)$	$G_3(3)$	...	$G_s(3)$
4	$G_1(4)$	$G_2(4)$	$G_3(4)$	...	$G_s(4)$
5	$G_1(5)$	$G_2(5)$	$G_3(5)$	...	$G_s(5)$
6	$G_1(6)$	$G_2(6)$	$C_3(6)$	...	$C_s(6)$
7	$G_1(7)$	$G_2(7)$	$G_3(7)$	...	$G_s(7)$
...	...	...	...	...	...
...	...	...	...	...	...
n	$G_1(n)$	$G_2(n)$	$G_3(n)$	...	$G_s(n)$

columns 2 to N and rows 2 to n+1, and starting from the element g(2,2), are calculated taking the element that is above the element to be calculated and multiplying it by x_j, where j is the column. The result of the preceding multiplication is added to the element to the left, and the result becomes the new element. See Fig. 6.15.

5. *G* Coefficient Results - When the calculation process is complete, the *G* coefficients are the elements of last column, that is, each element in the last column represents the *G* with a level of transactions given by its row number minus 1.

Formula (6.81) allows the calculation of $G_i(j)$ for all the processing stations. The values of the $G(j)$ are given by the last column of the previous table, or

$$G(j) = G_s(j) \tag{6.82}$$

3.3 Direct Calculation of G(j)

There is a need to directly determine the coefficients $G(j)$ [38]. To this end, taking into consideration Equations (6.72) and (6.73), we obtain

$$g(t) = (1 + x_1 t + x_1^2 t^2 + \cdots)(1 + x_2 t + x_2^2 t^2 + \cdots)$$
$$\cdots (1 + x_N t + x_N^2 t^2 + \cdots) =$$
$$= G(0) + G(1)t + G(2)t^2 + \cdots \tag{6.83}$$

Equation (6.83) can be written as follows

$$g(t) = \prod_{k=1}^{N} \frac{1}{1 - x_k t} \tag{6.84}$$

Two cases can occur when performing the partial fraction expansion:

• Different Service Times
• Multiple Service Times

3.3.1   Different Service Times

If all the service times are different, applying partial fraction expansion to Equation (6.84), we have

$$g(t) = \sum_{k=1}^{N} \frac{A_k}{1 - x_k t} \tag{6.85}$$

where the coefficients A_k are given by

$$A_k = \prod_{j=1/j \neq k}^{N} \frac{1}{1 - \dfrac{x_j}{x_k}} \tag{6.86}$$

In the above partial fraction expansion, we assume that $x_i \neq x_j$ for $i \neq j$. If this assumption does not hold, the above computation is somewhat more complicated, as will be discussed for the case of multiple service times.

Therefore, taking into consideration Equation (6.85), we can write Equation (6.86) as follows:

$$g(t) = \sum_{k=1}^{N} [A_k \sum_{j=0}^{\infty} (x_k t)^j]$$
$$= \sum_{j=0}^{\infty} [\sum_{k=1}^{N} A_k x_k^j] t^j \tag{6.87}$$

Then we write

$$G(j) = A_1 x_1^j + A_2 x_2^j + \cdots + A_N x_N^j \tag{6.88}$$

Thus, we have the result that $G(j)$ is a linear function of the jth power of $x_1, x_2, \ldots, x_N$.

3.3.2   Multiple Service Times

We review first the partial fraction expansion methodology for the case of multiple poles. We must remember that poles are the roots of the denominator of the expression to be expanded.

If we have multiple poles, we can write the polynomial fraction as follows:

$$\frac{A(s)}{B(s)} = \frac{A(s)}{(s-s_1)^{m_1}(s-s_2)^{m_2}\ldots(s-s_j)^{m_j}\ldots(s-s_n)^{m_n}} \qquad (6.89)$$

where the pole s_1 is of order m_1, the pole s_2 is of order m_2, and so on. The fraction $A(s)/B(s)$ can be resolved into a sum of partial fractions of the form

$$\frac{K_{j1}}{(s-s_j)^{m_j}}, \quad \frac{K_{j2}}{(s-s_j)^{m_j-1}}, \quad \ldots, \quad \frac{K_{jm_j}}{s-s_j} \qquad (6.90)$$

in which the K's are constants yet to be determined. Thus, the expansion of Equation (6.89) is

$$\frac{A(s)}{B(s)} = \frac{K_{11}}{(s-s_1)^{m_1}} + \frac{K_{12}}{(s-s_1)^{m_1-1}} + \cdots + \frac{K_{1m_1}}{s-s_1} + \cdots +$$

$$\frac{K_{k1}}{(s-s_k)^{m_k}} + \frac{K_{k2}}{(s-s_k)^{m_k-1}} + \cdots + \frac{K_{km_k}}{s-s_k} + \cdots +$$

$$\frac{K_{n1}}{(s-s_n)^{m_n}} + \frac{K_{n2}}{(s-s_n)^{m_n-1}} + \cdots + \frac{K_{nm_n}}{s-s_n} \qquad (6.91)$$

Therefore, we can write

$$\frac{A(s)}{B(s)} = \sum_{k=1}^{n}\sum_{j=1}^{m_j}\frac{K_{kj}}{(s-s_k)^{m_k-j+1}} \qquad (6.92)$$

where the K_{kj} is given by

$$K_{kj} = \frac{1}{(j-1)!}\left[\frac{d^{j-1}}{ds^{j-1}}\frac{(s-s_k)^{m_k}A(s)}{B(s)}\right]_{s=s_k} \qquad (6.93)$$

Example 6.2 - Let us expand in partial fractions the polynomial
fraction given below

$$\frac{A(s)}{B(s)} = s + \frac{3}{s^3 + 6s^2 + 7s + 4}$$

The roots of $B(s)$ are $s_{1,2} = -1$, which is a pole of second order, and
$s_3 = 4$. Therefore, we can write

$$\frac{A(s)}{B(s)} = \frac{K_{11}}{(s + 1)^2} + \frac{K_{12}}{s + 1} + \frac{K_2}{s + 4}$$

Applying Equation (6.91), we have:

$$K_{11} = \left[\frac{s + 3}{s + 4} \right]_{s = -1} = \frac{2}{3}$$

$$K_{12} = \left[\frac{d}{ds} \frac{s + 3}{s + 4} \right]_{s = -1} = \left[\frac{3(s + 4) - 4(s + 3)}{(s + 4)^2} \right]_{s = -1} = \frac{1}{9}$$

and

$$K_2 = \left[\frac{s + 3}{(s + 1)^2} \right]_{s = -4} = -\frac{1}{9}$$

Example 6.3 - Let us consider a local computer network consisting of
four processing nodes with service times given by

$$x_1 = 3.25$$
$$x_2 = 4.75$$
$$x_3 = 4.75$$
$$x_4 = 5.30$$

We wish to calculate the G coefficients for performance determination
purposes. The generating function of the network is given by:

$$g(t) = \frac{1}{1 - x_1 t} \frac{1}{(1 - x_2 t)^2} \frac{1}{1 - x_4 t}$$

If we apply partial fraction expansion, the above equation can be written
as

$$g(t) = \frac{K_1}{1-x_1 t} + \frac{K_{21}}{(1-x_2 t)^2} + \frac{K_{22}}{1-x_2 t} + \frac{K_4}{1-x_4 t}$$

where the coefficients are given by:

$$K_1 = \left[\frac{1}{(1-x_2 t)^2 (1-x_4 t)} \right]_{t=\frac{1}{x_1}}$$

$$K_{21} = \left[\frac{1}{(1-x_1 t)(1-x_4)} \right]_{t=\frac{1}{x_2}}$$

$$K_{22} = \left[\frac{d}{dt} \frac{1}{(1-x_1 t)(1-x_4 t)} \right]_{t=\frac{1}{x_2}}$$

$$K_4 = \left[\frac{1}{(1-x_1 t)(1-x_2 t)^2} \right]_{t=\frac{1}{x_4}}$$

In our case the coefficients are:

$$K_1 = \frac{1}{(1-\frac{4.75}{3.25})^2 (1-\frac{5.3}{3.25})} = 7.44$$

$$K_{21} = \frac{1}{1-\frac{3.25}{4.75})(1-\frac{5.3}{4.75})} = 236.40$$

$$K_{22} = \frac{3.25(1-\frac{5.3}{4.75}) + 5.3(1-\frac{3.25}{4.75})}{(1-\frac{3.25}{4.75})^2 (1-\frac{5.3}{4.75})^2} = 1533.89$$

$$K_4 = \frac{1}{(1-\frac{3.25}{5.3})^2 (1-\frac{4.75}{5.3})^2} = 240.09$$

The coefficient of $G(j)$ is given by

$$G(j) = K_1 x_1{}^j + K_{21} C_{2j} x_2{}^j + K_{22} x_2{}^j + K_4 x_4{}^j$$

which can be written as follows

$$G(j) = K_1 x_1{}^j + K'_2 x_2{}^j + K_4 x_4{}^j$$

where

$$K'_2 = K_{21} C_{2j} + K_{22}$$

Here $C_{2j} = 2$; therefore,

$$K'_2 = K_{21} 2 + K_{22}$$

3.3.3   Determination of C_{nj}

Let us now discuss how to obtain the coefficient C_{nj}. In doing this, we consider the following power series

$$(1 + xt + x^2t^2 + x^3t^3 + \cdots)^n =$$

$$1 + n\,xt + [\frac{n+(n+1)}{2}] x^2t^2 + [\frac{n+(n+1)+(n+2)}{6}] x^3t^3 + \cdots$$

$$+ [\frac{n+(n+1)+(n+2)+ \cdots +(n+j-1)}{j!}] x^j t^j + \cdots \qquad (6.94)$$

We can write

$$C_{nj} = \begin{bmatrix} n+j-1 \\ j \end{bmatrix} = \frac{(n+j-1)!}{j!(n-1)!} \qquad (6.95)$$

To obtain the above equation, we consider the following power series

$$f(z) = (1 + z + z^2 + z^3 + z^4 + \cdots)^n = \frac{1}{(1-z)^n} \qquad (6.96)$$

Assume that we want to develop the previous function $f(z)$ as a Taylor series that can be written as

$$f(z) = \frac{1}{(1-z)^n} = \sum_{k=0}^{\infty} \frac{f^{(k)}(0)}{k!} z^k \qquad (6.97)$$

We calculate the successive derivatives

$$f'(z) = \frac{n}{(1-z)^{n+1}}$$

$$f''(z) = \frac{n(n+1)}{(1-z)^{n+2}} \qquad (6.98)$$

and so on. For the kth derivative, we write

$$f^{(k)}(z) = \frac{n(n+1)....(n+k-1)}{(1-z)^{n+k}} \qquad (6.99)$$

for $z=0$ we have

$$f^{(k)}(0) = n(n+1) \; .... \; (n+k-1) \tag{6.100}$$

Multiplying and dividing Equation (6.100) by $(n-1)!$, we have

$$f^{(k)}(0) = \frac{(n-1)! \; n(n+1)....(n+k-1)}{(n-1)!} \tag{6.101}$$

This equation can be written as follows

$$f^{(k)}(0) = \frac{(n+k-1)!}{(n-1)!} \tag{6.102}$$

Therefore, Equation (6.97) becomes

$$f(z) = \sum_{k=0}^{\infty} \frac{(n+k-1)!}{k!(n-1)!} \, z^k \tag{6.103}$$

Now, performing the necessary change of variables in Equation (6.103), we obtain Equation (6.97).

3.4   General Solution for One Class of Transactions

Assume that in Equation (6.99) we have k multiple roots, each of order m_q, for $q=1,2,3,...,k$. Ordering the roots such that the multiple roots are at the end, we can write

$$G(j) = K_1 \, x_1^j + K_2 \, x_2^j + \cdots + K_{n-k} \, x_{n-k}^j \tag{6.104}$$

$$+ \sum_{q=n-k+1}^{n} \sum_{l=1}^{m_q} C_{m_q j} \, K_{ql} \, x_q^j$$

Equation (6.104) gives us the value of $G(j)$ for any particular problem. This equation can be written as follows

$$G(j) = K_1 \, x_1{}^j + K_2 \, x_2{}^j + \cdots + K_{n-k} \, x_{n-k}{}^j$$

$$+ \sum_{q=n-k+1}^{n} \sum_{l=1}^{m_q} \frac{(mm_q + j - 1)!}{j! \, (m_q - 1)!} \, K_{ql} \, x_q{}^j \tag{6.105}$$

where the coefficients are given by

$$K_j =$$

$$\left[\frac{1-x_j t}{(1-x_1 t) \cdots (1-x_{n-k} t)(1-x_{n-k+1} t)^{m_n - k + 1} \cdots (1-x_n t)^{m_n}} \right]_{t=\frac{1}{x_j}} \tag{6.106}$$

$$K_{ij} = \frac{1}{(j-1)!}$$

$$\left[\frac{d^{j-1}}{dt^{j-1}} \ \frac{(1-x_j t)^{m_i}}{(1-x_1 t) \ \cdots \ (1-x_{n-k} t)(1-x_{n-k+1} t)^{m_n-k+1} \ \cdots \ (1-x_n t)^{m_n}} \right]_{t=\frac{1}{x_j}}$$

(6.107)

3.4.1 Performance Indices

The performance indices of the network system can be obtained directly from the G coefficients of the generating function of the network. We will be analyzing the following performance indices:

- Device Utilization $U_i(n)$
- Overlapping Utilization $U_{ij}(n)$
- Queue Length $Q_i(n)$
- Residence Time $R(n)$
- Throughput $T(n)$

3.4.2 Device Utilization

Assuming that the generating function of the processing network is given by

$$
\begin{aligned}
g(t) = & (\ 1 \ + \ x_1 t \ + \ x_1^2 t^2 \ + \ x_1^3 t^3 \ + \ ...) \\
& (\ 1 \ + \ x_2 t \ + \ x_2^2 t^2 \ + \ x_2^3 t^3 \ + \ \cdots \ ) \ \cdots \\
& (\ 1 \ + \ x_i t \ + \ x_i^2 t^2 \ + \ x_i^3 t^3 \ + \ \cdots \ ) \ \cdots \\
& (\ 1 \ + \ x_N t \ + \ x_N^2 t^2 \ + \ x_N^3 t^3 \ + \ \cdots \ ) \\
= & \ G(0) \ + \ G(1)t \ + \ G(2)t^2 \ + \ G(3)t^3 \ + \ \cdots
\end{aligned}
$$

(6.108)

we subtract 1 from the power series which represents device i; this results in a new generating function which assumes that device i has at least one transaction processing, or in other words, that device i is busy. Let us designate this function by $h_i(t)$

$$h_i(t) = g(t) \ \frac{x_i(t)-1}{x_i(t)}$$

(6.109)

It is easy to see that $h_i(t)$ is obtained by multiplying $g(t)$ by $[x_i(t)-1]$ and dividing by $x_i(t)$. The above equation can be written as

$$h_i(t) = g(t) \left[1 - \frac{1}{x_i(t)} \right] \tag{6.110}$$

But since we know that

$$x_i(t) = 1 + x_i t + x_i^2 t^2 + x_i^3 t^3 + \cdots = \frac{1}{1 - x_i t} \tag{6.111}$$

we can write

$$h_i(t) = g(t)\, x_i\, t \tag{6.112}$$

The above equation can be expanded as a power series in $g(t)$; thus,

$$h_i(t) = H_i(0) + H_i(1)t + H_i(2)t^2 + H_i(3)t^3 + \cdots \tag{6.113}$$

The value of $H_i(0) = 0$, as we expect. The utilization of device i is given by

$$U_i(n) = \frac{H(n)}{G(n)} = x_i \frac{G(n-1)}{G(n)} \tag{6.114}$$

Now

$$h'_i(t) = g(t) \frac{x_i(t) - 1 - x_i t}{x_i(t)} \tag{6.115}$$

In the above equation we are subtracting from $g(t)$ the first two terms of $x_i(t)$, which represents device i with one transaction processing and the other transaction waiting in the device queue. We can write the above equation as follows

$$h'_i(t) = g(t) \left[1 - \frac{1 + x_i\, t}{x_i(t)} \right] = g(t)\, [1 - (1 + x_i t)(1 - x_i t)]$$

$$= g(t)\, x_i^2\, t^2 \tag{6.116}$$

Therefore, we can write

$$U_i^2 = x_i^{\,2} \frac{G(n-2)}{G(n)} \tag{6.117}$$

and generalizing, we obtain

$$U_i^k = x_i^k \frac{G(n-k)}{G(n)} \tag{6.118}$$

3.4.3 Overlapping Utilization

Similarly, we can define

$$h_{ij}(t) = g(t) \frac{x_i(t) - 1}{x_i(t)} \frac{x_j(t) - 1}{x_j(t)} \tag{6.119}$$

Then,

$$h_{ij}(t) = g(t) \, x_i \, x_j \, t^2 \tag{6.120}$$

After some manipulations we obtain

$$U_{ij}(n) = x_i x_j \frac{G(n-2)}{G(n)} \tag{6.121}$$

which represents the utilization of devices i and j at the same time. This performance index plays an important role in device scheduling.

After multiplying and dividing Equation (6.121) by $G(n-1)$, we obtain

$$U_{ij}(n) = x_i \, x_j \frac{G(n-2)}{G(n)} \frac{G(n-1)}{G(n-1)} \tag{6.122}$$

which can be written as follows

$$U_{ij}(n) = x_i \frac{G(n-1)}{G(n)} x_j \frac{G(n-2)}{G(n-1)} \tag{6.123}$$

and if we take into consideration Equation (6.114), Equation (6.123) becomes

$$U_{ij}(n) = U_i(n) \, U_j(n-1) \tag{6.124}$$

Swapping the positions of x_i and x_j in Equation (6.122), we can write

$$U_{ij}(n) = U_i(n-1) \, U_j(n) \tag{6.125}$$

Equations (6.124) and (6.125) allow us to write

$$U_i(n) \, U_j(n-1) = U_i(n-1) \, U_j(n) \tag{6.126}$$

3.4.4 Queue Length

The queue length of device i is given by

$$Q_i(n) = P_i(n_i \geqslant 1) + P_i(n_i \geqslant 2) + P_i(n_i \geqslant 3) + \cdots \tag{6.127}$$

or,

$$Q_i(n) = \sum_{k=1}^{\infty} x_i^k \frac{G(n-k)}{G(n)} \tag{6.128}$$

Since we know that $n = \sum\limits_{i=1}^{N} n_i$, we have

$$n = \sum_{i=1}^{N} Q_i(n) = \frac{1}{G(n)} \sum_{i=1}^{N} \sum_{j=1}^{n} x_i^j \, G(n-j) \qquad (6.129)$$

or,

$$G(n) = \frac{1}{n} \sum_{j=1}^{n} [\sum_{i=1}^{N} x_i^j \,] \, G(n-j) \qquad (6.130)$$

which can serve as an alternative algorithm for calculating $G(j)$.

3.4.5   Residence Time

To obtain the residence time $R(n)$, which represents the time that a transaction remains in the processing network, we need to look at an arbitrary processor i and count the times r that the transaction needs to visit the processor before leaving the network. Given processor i and the number of times r, we can write

$$\frac{U_i(n)}{n} \, R(n) = \tau_i \, r$$

The utilization of processor i per transaction is given by $U_i(n)/n$; therefore, multiplying this value for $R(n)$ represents the time that the transaction was in the processing network. This value is equal to $\tau_i \, r$.

The residence time is given by

$$R(n) = \frac{rn}{\rho_i} \frac{G(n)}{G(n-1)} \qquad (6.131)$$

3.4.6   Throughput

The throughput, which is the inverse of the residence time multiplied by n, is given by

$$T(n) = \frac{\rho_i}{r} \frac{G(n-1)}{G(n)} \qquad (6.132)$$

3.5   Direct Calculation of Performance Indices

We already studied the direct calculation of the generating function coefficients, assuming that all roots are different. We now want to write the performance indices in closed-form. In this case we can write:

$$G(j) = A_1 x_1{}^j + A_2 x_2{}^j + \cdots + A_N x_N{}^j \qquad (6.133)$$

Thus, $G(j)$ is a linear function of the jth power of $x_1, x_2, \ldots, x_N$. The utilization of the ith processing station is given by

$$U_i(n) = x_i \frac{A_1 x_1^{n-1} + A_2 x_2^{n-1} + \cdots + A_N x_N^{n-1}}{A_1 x_1^n + A_2 x_2^n + \cdots + A_N x_N^n} \qquad (6.134)$$

The processing station overlap between stations i and j is given by

$$U_{ij}(n) = x_i x_j \frac{A_1 x_1^{n-2} + A_2 x_2^{n-2} + \cdots + A_N x_N^{n-2}}{A_1 x_1^n + A_2 x_2^n + \cdots + A_N x_N^n} \qquad (6.135)$$

The queue length at processing station i is given by

$$Q_i(n) = \frac{\sum_{j=1}^{n} [A_1 x_1^{n-j} + A_2 x_2^{n-j} + \cdots + A_N x_N^{n-j}]}{A_1 x_1^n + A_2 x_2^n + \cdots + A_N x_N^n} \qquad (6.136)$$

The residence time is given by

$$R(n) = \frac{rn}{\rho_i} \frac{A_1 x_1^n + A_2 x_2^n + \cdots + A_N x_N^n}{A_1 x_1^{n-1} + A_2 x_2^{n-1} + \cdots + A_N x_N^{n-1}} \qquad (6.137)$$

and the throughput is given by

$$T(n) = \frac{\rho_i}{r} \frac{A_1 x_1^{n-1} + A_2 x_2^{n-1} + \cdots + A_N x_N^{n-1}}{A_1 x_1^n + A_2 x_2^n + \cdots + A_N x_N^n} \qquad (6.138)$$

The formulas needed for performance measurement calculations in the case of a single class of transactions, when all the roots are different, are shown in Table 6.2.

We now present an application of the preceding material, particularly to analyze the results obtained with the purpose of introducing linear models for performance determinations.

Example 6.4 - Consider the computer network shown in Fig. 6.14. We want to write the generating function of the network and in particular to obtain a general formula for G coefficients.

The service times are

$$\tau_0 = 3.20$$
$$\tau_1 = 3.85$$
$$\tau_2 = 4.21$$
$$\tau_3 = 3.26$$
$$\tau_4 = 3.72$$

Table 6.2 Formulas for Single Class of Transaction

Definition	Variable	Formula
Utilization	$U_i(n)$	$U_i(n) = x_i \dfrac{A_1 x_1^{n-1} + A_2 x_2^{n-1} + \cdots + A_N x_N^{n-1}}{A_1 x_1^{n} + A_2 x_2^{n} + \cdots + A_N x_N^{n}}$
Overlapping Utilization	$U_{ij}(n)$	$U_{ij}(n) = x_i x_j \dfrac{A_1 x_1^{n-2} + A_2 x_2^{n-2} + \cdots + A_N x_N^{n-2}}{A_1 x_1^{n} + A_2 x_2^{n} + \cdots + A_N x_N^{n}}$
Queue Length	$Q_i(n)$	$Q_i(n) = \dfrac{\sum\limits_{j=1}^{n} [A_1 x_1^{n-j} + A_2 x_2^{n-j} + \cdots + A_N x_N^{n-j}]}{A_1 x_1^{n} + A_2 x_2^{n} + \cdots + A_N x_N^{n}}$
Residence Time	$R(n)$	$R(n) = \dfrac{rn}{\rho_i} \dfrac{A_1 x_1^{n} + A_2 x_2^{n} + \cdots + A_N x_N^{n}}{A_1 x_1^{n-1} + A_2 x_2^{n-1} + \cdots + A_N x_N^{n-1}}$
Throughput	$T(n)$	$T(n) = \dfrac{\rho_i}{r} \dfrac{A_1 x_1^{n-1} + A_2 x_2^{n-1} + \cdots + A_N x_N^{n-1}}{A_1 x_1^{n} + A_2 x_2^{n} + \cdots + A_N x_N^{n}}$

To calculate the transfer probabilities for full capacity, we let

$$n_1 \tau_1 = n_2 \tau_2 = n_3 \tau_3 = n_4 \tau_4$$

Assuming $n_1 = 1$, we obtain

$$n_2 = \frac{n_1 \tau_1}{\tau_2} = \frac{3.85}{4.21} = 0.9144$$

$$n_3 = \frac{n_1 \tau_1}{\tau_3} = \frac{3.85}{3.26} = 1.1809$$

$$n_4 = \frac{n_1 \tau_1}{\tau_4} = \frac{3.85}{3.72} = 1.0349$$

The total average number of processing transactions is

$$\Delta = n_1 + n_2 + n_3 + n_4$$
$$= 1 + 0.9144 + 1.1809 + 1.0349 = 4.1302$$

The transfer probabilities are

$$p_{01} = \frac{n_1}{\Delta} = \frac{1}{4.1302} = 0.2421$$

$$p_{02} = \frac{n_2}{\Delta} = \frac{0.9144}{4.1302} = 0.2213$$

$$p_{03} = \frac{n_3}{\Delta} = \frac{1.1809}{4.1302} = 0.2859$$

$$p_{04} = \frac{n_4}{\Delta} = \frac{1.0349}{4.1302} = 0.2505$$

We obtain the relative frequencies by solving the following system of equations

$$\sum_{i=1}^{4} \rho_i \, p_{ji} = \rho j \qquad j = 1,2,3,4$$

Therefore, we have

$$\rho_1 = 0.2421 \, \rho_1 + 0.2213 \, \rho_2 + 0.2859 \, \rho_3 + 0.2505 \, \rho_4$$

$$\rho_2 = 0.1421 \, \rho_1 + 0.2213 \, \rho_2 + 0.2859 \, \rho_3 + 0.2505 \, \rho_4$$

$$\rho_3 = 0.2421 \, \rho_1 + 0.2213 \, \rho_2 + 0.2859 \, \rho_3 + 0.2505 \, \rho_4$$

$$\rho_4 = 0.2421 \, \rho_1 + 0.2213 \, \rho_2 + 0.2859 \, \rho_3 + 0.2505 \, \rho_4$$

Which can be written as follows

$$-0.7579 \, \rho_1 + 0.2213 \, \rho_2 + 0.2859 \, \rho_3 + 0.2505 \, \rho_4 = 0$$

$$0.1421 \, \rho_1 - 0.7787 \, \rho_2 + 0.2859 \, \rho_3 + 0.2505 \, \rho_4 = 0$$

$$0.2421 \, \rho_1 + 0.2213 \, \rho_2 - 0.7141 \, \rho_3 + 0.2505 \, \rho_4 = 0$$

$$0.2421 \, \rho_1 + 0.2213 \, \rho_2 + 0.2859 \, \rho_3 - 0.7495 \, \rho_4 = 0$$

From inspection of the above equations, the solution is obtained when all the ρ's are equal. Thus,

$$\rho_1 = \rho_2 = \rho_3 = \rho_4$$

Therefore, we choose the values $\rho_i = 1.0$ for $i = 1,2,3,4$. The normalized service times are

$$x_i = \rho_i \, \tau_i$$

Since $\rho_i = 1$, we see that the concept of maximum capacity makes the normalized service variables equal to the service processing times.

Table 6.3 G_i Coefficients

n	G_0	G_1	G_2	G_3	G_4
0	1.00	1.00	1.00	1.00	1.00
1	3.20	7.05	11.26	14.52	18.24
2	10.24	37.38	84.79	132.12	199.98
3	32.76	176.69	533.64	964.36	708.27
4	104.85	705.12	3031.76	6175.58	12,530.35
5	335.54	3358.24	16,121.95	36,254.35	82,867.25
6	1073.74	14,002.98	81,876.38	200,065.60	508,331.80
7	3435.97	57,347.43	402,047.00	1,054,261.00	2,945,255.00

Table 6.3 presents the values of the G_i. The $G(j)$ coefficients are those indicated in the right column of Table 6.2.

We now use partial fraction expansion in the generating function of the processing network. Thus,

$$g(t) = (1 + x_0 t + x_0^2 t^2 + x_0^3 t^3 + \cdots)$$
$$(1 + x_1 t + x_1^2 t^2 + x_1^3 t^3 + \cdots) \cdots$$
$$(1 + x_4 t + x_4^2 t^2 + x_4^3 t^3 + \cdots)$$
$$= \frac{A_0}{1-x_0 t} + \frac{A_1}{1-x_1 t} + \cdots + \frac{A_4}{1-x_4 t}$$

Then we know that

$$G(j) = A_0 x_0^j t^j + A_1 x_1^j t^j + \cdots + A_4 x_4^j t^j$$

The values of the coefficients are:

$$A_0 = \frac{1}{1-x_1 t} \frac{1}{1-x_2 t} \frac{1}{1-x_3 t} \frac{1}{1-x_4 t} \Big|_{t=\frac{1}{x_0}}$$

$$= \frac{1}{1-\dfrac{3.85}{3.2}} \frac{1}{1-\dfrac{4.21}{3.2}} \frac{1}{1-\dfrac{3.26}{3.2}} \frac{1}{1-\dfrac{3.72}{3.2}}$$

$$= 5119.3$$

$$A_1 = -12241.41$$

$$A_2 = 1856.03$$

$$A_3 = -7301.05$$

$$A_4 = 12568.14$$

As we expected, because $G(0) = 1.0$, we have

$$A_0 + A_1 + A_2 + A_3 + A_4 =$$

$$5119.3 - 12241.41 + 1856.03 - 7301.05 + 12568.14 = 1.0$$

Therefore, the equation for the $G(j)$ coefficients is

$$G(j) = 5119.3 \ (3.2)^j - 12241.41 \ (3.85)^j + 1856.03 \ (4.21)^j$$
$$- 7301.05 \ (3.26)^j + 12568.14 \ (3.72)^j$$

The processor utilization values for n from 2 to 20 are given in Table 6.4.

Table 6.4 Processor Utilization

n	$U_0(n)$	$U_1(n)$	$U_2(n)$	$U_3(n)$	$U_4(n)$
2	29.2	35.1	38.4	29.7	33.9
3	37.5	45.1	49.3	38.2	43.5
4	37.5	52.5	57.4	44.4	52.7
5	48.4	58.2	63.7	49.3	56.3
6	52.2	62.8	68.6	53.1	60.6
7	55.2	66.4	72.7	56.3	64.2
8	57.8	69.5	76.0	58.8	67.1
9	59.9	72.0	78.8	61.0	69.6
10	61.7	74.2	81.1	62.8	71.7
11	63.2	76.0	83.2	64.4	73.5
12	64.5	77.6	84.9	65.7	75.0
13	65.7	79.0	86.4	66.9	76.4
14	66.7	80.2	87.7	67.9	77.5
15	67.6	81.3	88.9	68.8	78.6
16	68.4	82.3	89.9	69.7	79.5
17	69.1	83.1	90.9	70.4	80.3
18	69.7	83.9	91.7	71.0	81.0
19	70.3	84.5	92.4	71.6	81.7

We can see that the utilization of any processor, for any number of transactions in the system, can be calculated directly, i.e., the utilization of processor 2 for $n = 10$ is given by

$$U_2(10) = 74.2$$

And the overlapping between processors 1 and 2 is given by

$$U_{12}(10) = 0.04$$

The base overlapping utilization is defined by $U_2(n)$, which in our case refers to I/O2. Therefore, the associated overlapping utilizations are given in Table 6.5. The residence time and throughput appear in Table 6.6, and the q-lengths are given in Table 6.7.

Observing the processor utilization figures in the preceding table, we see that they increase when the number of transactions in the processing network increases, which is the expected behavior. We now proceed to analyze the following data

Table 6.5 Overlapping Utilization

n	$U_{02}(n)$	$U_{12}(n)$	$U_{32}(n)$	$U_{42}(n)$
2	6.7	8.1	6.9	7.8
3	14.4	17.3	14.7	16.7
4	21.5	25.9	21.9	25.0
5	27.8	33.4	28.3	32.3
6	33.2	40.0	33.8	38.6
7	37.9	45.6	38.6	44.1
8	42.0	50.5	42.8	48.8
9	45.5	54.7	46.4	52.9
10	48.6	58.4	49.5	56.5
11	51.3	61.7	52.2	59.6
12	53.7	64.6	54.7	62.4
13	55.8	67.1	56.8	64.8
14	57.6	69.3	58.7	67.0
15	59.3	71.3	60.4	68.9
16	60.8	73.1	61.9	70.7
17	62.1	74.7	63.3	72.2
18	63.3	76.2	64.5	73.6
19	64.4	77.5	65.6	74.9

Table 6.6 Residence Time and Throughput

n	Residence Time	Throughput
2	36.5	197.4
3	32.9	328.4
4	34.2	421.4
5	36.7	490.8
6	39.7	544.4
7	42.9	586.9
8	46.4	621.3
9	49.9	649.8
10	53.4	673.6
11	57.1	693.8
12	60.8	711.0
13	64.5	726.0
14	68.2	738.9
15	72.0	750.3
16	75.8	760.3
17	75.8	769.2
18	83.4	777.0
19	87.2	784.1

Table 6.7 Q-length at each Processor

n	CPU	I/O1	I/O2	I/O3	I/O4
1	.175	.211	.231	.179	.204
2	.343	.425	.473	.350	.409
3	.503	.642	.726	.515	.613
4	.656	.862	.991	.673	.818
5	.801	1.084	1.267	.825	1.023
6	.940	1.308	1.556	.970	1.227
7	1.071	1.534	1.857	1.108	1.430
8	1.196	1.761	2.171	1.241	1.631
9	1.315	1.989	2.498	1.367	1.832
10	1.428	2.217	2.838	1.487	2.030
11	1.534	2.447	3.191	1.601	2.226
12	1.635	2.676	3.558	1.710	2.420
13	1.731	2.905	3.939	1.813	2.612
14	1.821	3.133	4.334	1.912	2.800
15	1.907	3.361	4.742	2.005	2.985
16	1.987	3.587	5.165	2.093	3.168
17	2.063	3.812	5.602	2.176	3.346
18	2.135	4.035	6.054	2.255	3.521
19	2.202	4.256	6.520	2.330	3.693

1. **Maximum Processor Utilization** - Setting a practical limit.
2. **Linear Model Approximation** - Convenient approximation for mixed problem determination.

Maximum Processor Utilization - Since we consider a **closed network**, the total number of transactions in the processing network is as indicated in column 1, Table 6.4, so that for 5 transactions in the network, we have

$$U_0(5) = 48.4\%$$

$$U_1(5) = 58.2\%$$

$$U_2(5) = 63.7\%$$

$$U_3(5) = 49.3\%$$

$$U_4(5) = 56.3\%$$

We also observe that the processor which brings the network into saturation is the one that has the maximum utilization, in this case, processor number 2. We also observe that this particular processor has the highest utilization for any number of transactions in the network. Comparing the utilization with the number of transactions, we see that the increase in utilization for the same increment in the number of transactions is almost constant, until the utilization reaches over 90%, which defines for us the point at which the processing network becomes saturated. Beyond this point, the rate of increase in the utilization decreases considerably, building up the queue length for the processor. Only when $n \to \infty$, the utilization of the processor becomes 100%, or in our example

$$U_2(\infty) = 100.0\%$$

Therefore, if we want to keep the network operating outside the saturation region, we need to define a criteria for selecting the maximum utilization. Once this criteria is defined, we can find the total number of transactions in the processing network that maintains operation within the unsaturated region. In general for local computer networks a good criterion for the maximum utilization is around 93%. Whichever criterion is used, the maximum number of transactions in the network defines its **capacity**.

3.6 Linear Model Approximation

For some types of applications it is convenient to represent the function $U_j(n)$ by a linear model, but precaution must be exercised selecting the linear range of n values. For example, observing the data presented in

Table 6.4, we see that for processing device 2, in the range of 2-20 transactions, we can use least squares to obtain the following linear equation:

$$U_2(n) = 49.27 + 2.58 \times n$$

If we use this formula to calculate $U_2(2)$, we obtain a value with an error of 41.77%, and for $U_2(n)$ an error of 15.66%, and so on. These large errors occur because the utilization in the range 2-20 does not behave linearly. On the other hand, if we chose the interval 5-20, we can represent the utilization of device 2 by means of the following linear formula:

$$U_2(n) = 60.602 + 1.801 n$$

The values of 60.602 and 1.801 have been obtained using the Least-Squares Method, for the range of n from 5-20. Table 6.8 presents the errors obtained using the linear approximation.

Observing the preceding table, we see that the linear model is a good representation of the utilization of processing device 2 in the range 5-20. Now, if we apply this model for $n=25$, we have

$$U_2(25) = 60.602 + 1.801 \ x \ 25 = 105.62\%$$

Table 6.8   Absolute and Percent Errors

n	$U_2(n)$	$U_{2c}(n)$	Abs. Error	% Errors
5	63.7	69.612	5.912	9.28
6	68.6	71.474	2.814	4.10
7	72.7	73.215	0.515	0.70
8	76.0	75.017	-0.983	-1.29
9	78.8	76.819	-1.981	-2.51
10	81.1	78.621	-2.479	-3.05
11	83.2	80.422	-2.778	-3.33
12	84.9	82.224	-2.676	-3.15
13	86.4	84.026	-2.374	-2.13
14	87.7	85.828	-1.872	-2.13
15	88.9	87.629	-1.271	-1.42
16	89.9	89.431	-0.469	-0.52
17	90.9	91.233	0.333	0.36
18	91.7	93.035	1.335	1.45
19	92.4	94.836	2.436	2.63
20	93.1	96.638	3.538	3.80

which is incorrect, since the linear model representation is only valid for the range 5-20. We observe that if our criterion is to assume that the system is saturated for $U_2(n)$ equal to or greater than 93.00%, from Table 6.4 we see that the maximum number of transactions in the system will be $n = 19$. Using our linear model, we have

$$60.602 + 1.801\,n = 93$$

or,

$$n = \frac{93 - 60.602}{1.801} = 17.98 \ \ or \ \ n \approx 18$$

Using the linear model for 100%, we have

$$n = \frac{100.0 - 60.602}{1.801} = 21.87 \ \ or \ \approx 22$$

It is obvious that the criterion for choosing the value of $U_2(n) = 93.0$ has to do with the start of saturation; the exact point at which this occurs is completely arbitrary. Therefore, it is convenient when using linear models to choose $U_2(n) = 100.0\%$.

4. CASE II: TWO CLASSES OF TRANSACTIONS

For the case of two different classes of transactions in the processing network, we define the state of the network as

$$S = (n_{11}, n_{12}, n_{21}, n_{22}, \cdots, n_{N1}, n_{N2}) \tag{6.139}$$

where n_{ij} denotes processor i and transaction class j. The total number of transactions in the processing network is

$$n = \sum_{i=1}^{m} \sum_{j=1}^{2} n_{ij} \tag{6.140}$$

The generating function of the processing network is given by

$$
\begin{aligned}
g(t,u) = \ & [1 + x_{11}t + x_{12}u + (x_{11}t + x_{12}u)^2 + \cdots] \\
& + [1 + x_{21}t + x_{22}u + (x_{21}t + x_{22}u)^2 + \cdots] \\
& + \cdots \\
& + [1 + x_{m1}t + x_{m2}u + (x_{m1} + x_{m2})^2 + \cdots] \quad (6.141)
\end{aligned}
$$

After all the transactions are considered, Equation (6.141) becomes

$$g(t,u) = G(0,0) + G(1,0)t + G(2,0)t^2 + \cdots + G(0,1)u$$
$$+ G(0,2)u^2 + \cdots + G(n_t,n_u)t^{n_t} u^{n_u} \quad (6.142)$$

4.1   Direct Calculation of G Coefficients

The generating function of the processing network when two transactions $(m=2)$ share the processing station is given by:

$$g(t,u) = \prod_{j=1}^{s} \frac{1}{1 - x_j t - y_j u} \quad (6.143)$$

Observing Equation (6.143), we see that we need to expand the product of factors into a summation, using for this purpose partial fraction expansion. This will be done in two steps. First we assume that one of the variables is constant, expanding with respect to the other. We then proceed assuming that the constant is now a variable, and that the variable is now a constant. If t is a constant, applying partial fraction expansion with respect to the variable u, we can write

$$\sum_{j=1}^{s} \frac{1}{(1 - x_j t) - y_j u} \prod_{i \neq j} \frac{1}{(1 - x_i t) - y_i \dfrac{(1 - x_i t)}{y_j}} \quad (6.144)$$

To obtain Equation (6.144), we note that the roots of the denominator are obtained as follows

$$1 - x_j t - y_j u = 0$$

Then,

$$u = \frac{1 - x_j t}{y_j}$$

Now, Equation (6.144) can be written as follows

$$\sum_{j=1}^{s} \frac{1}{1 - x_j t - y_j u} \prod_{i \neq j} \frac{y_j}{\delta_{ij} - \beta_{ij} t} \quad (6.145)$$

where δ_{ij} and β_{ij} are given by

$$\beta_{ij} = y_j x_i - y_i x_j$$
$$\delta_{ij} = y_j - y_i$$

Repeating the process of partial fractions, this time with respect to t, we obtain

$$\sum_{j=1}^{s} \frac{1}{1-x_j t - y_j u} \sum_{i=1/i\neq j}^{s} \frac{\alpha_{ij}}{\delta_{ij} - \beta_{ij} t} \tag{6.146}$$

where

$$\alpha_{ij} = \prod_{i=1/l\neq i/l\neq j}^{s} \frac{y_j}{\delta_{lj} - \beta_{lj}\dfrac{\delta_{ij}}{\beta_{ij}}} \tag{6.147}$$

Equation (6.146) can be written as

$$\sum_{j=1}^{s} \sum_{i=1/i\neq j}^{s} \frac{\alpha_{ij}/\delta_{ij}}{(1-x_j t - y_j u)(1-\gamma_{ij} t)} \tag{6.148}$$

$$\sum_{j=1}^{s} \sum_{i=1/i\neq j}^{s} \frac{\alpha_{ij}}{\delta_{ij}} \left[\sum_{n=0}^{\infty} \left[\sum_{n=0}^{\infty} (x_j t + y_j u)^n \right] \left[\sum_{m=0}^{\infty} (\gamma_{ij} t)^m \right] \right] \tag{6.149}$$

where

$$\gamma_{ij} = \frac{\beta_{ij}}{\delta_{ij}}$$

Then the generating function of the processing network is given by

$$g(t,u) =$$

$$\sum_{j=1}^{s} \sum_{i=1/i\neq j}^{s} \frac{\alpha_{ij}}{\delta_{ij}} \left[\sum_{n=0}^{\infty} \sum_{k=0}^{n} \binom{n}{k} (x_j^k y_j^{n-k}) t^k u^{n-k} \right] \left[\sum_{m=0}^{\infty} \gamma_{ij}^m t^m \right] \tag{6.150}$$

We can write the above equation as follows

$$g(t,u) =$$

$$\sum_{j=1}^{s} \sum_{i=1/i\neq j}^{s} \frac{\alpha_{ij}}{\delta_{ij}} \sum_{k=0}^{\infty} \left[\sum_{m=0}^{k} \sum_{l=0}^{\infty} \binom{l+m}{m} \gamma_{ij}^{k-m} x_j^m y_j^l u^l \right] t^k \tag{6.151}$$

Finally, if we switch the order of summation, the generating function of the processing network becomes

$$g(t,u) =$$

$$\sum_{k=0}^{\infty} \sum_{l=0}^{\infty} \left[\sum_{j=1}^{s} \sum_{i=1/i\neq j}^{s} \sum_{m=0}^{k} \frac{\alpha_{ij}}{\delta_{ij}} \binom{l+m}{m} \gamma_{ij}^{k-m} x_j^m y_j^l \right] u^l t^k \tag{6.152}$$

The coefficient $G(n_u, n_t)$ is given by

$$G(n_u, n_t) =$$

$$\sum_{j=1}^{s} \sum_{i=1/i \neq j}^{s} \sum_{m=0}^{n_t} \frac{\alpha_{ij}}{\delta_{ij}} \binom{n_u + m}{m} \gamma_{ij}^{n_t - m} x_j^m y_j^{n_u} \qquad (6.153)$$

The above equation can be obtained from Equation (6.152) with the following substitutions

$$l \Rightarrow n_u$$

$$k \Rightarrow n_t$$

The performance indices can be calculated from the G's, the coefficients of the generating function of the network. Table 6.9 presents the formulas to be applied in this case. The formulas in the table allow us to calculate the performance indices for class n_t. If we are interested in the performance for both classes together, we need to add the respective indices. As an example, for the utilization, we have

$$U(n_t, n_u) = U_t(n_t, n_u) + U_u(n_t, n_u) \qquad (6.154)$$

$$= \frac{x_{it}\, G(n_t - 1, n_u)}{G(n_t, n_u)} + \frac{x_{iu}\, G(n_t, n_u - 1)}{G(n_t, n_u)}$$

Table 6.9 Performance Indices Formulas for $m = 2$

Performance Index	Formula
Throughput	$T_t(n_t, n_u) = \dfrac{1}{k_t} \dfrac{G(n_t - 1, n_u)}{G(n_t, n_u)}$
Residence Time	$R_t(n_t, n_u) = k_t \dfrac{G(n_t, n_u)}{G(n_t - 1, n_u)}$
Queue Length	$Q_{it}(n_t, n_u) = \dfrac{1}{G(n_t, n_u)} \sum_{j=1}^{n} x_{it}^j G(n_t - j, n_u)$
Processing Station Utilization	$U_i(n_t, n_u) = \dfrac{x_{it} G(n_t - 1, n_u)}{G(n_t, n_u)}$
Processing Overlapping Utilization	$U_{ij}(n_t, n_u) = \dfrac{x_{it} x_{jt} G(n_t - 2, n_u)}{G(n_t, n_u)}$

and similarly for the other performance indices.

Several computer performance evaluation packages have been designed using Buzen's approach, such as BEST/1 [4] and BWQUE [17].

5. CASE III: MULTIPLE CLASSES OF TRANSACTIONS

In the previous discussion, we assumed one or two different classes of transactions. Now we would like to generalize to m different classes. Let us make the following definitions

$$\tau_{ik} = expected\ processing\ time\ at\ device\ i\ for\ class\ k$$

$$p_{ijk} = probability\ that\ transaction\ k\ goes\ device\ j\ after\ i$$

$$\rho_{ik} = relative\ frequency\ at\ device\ i\ for\ transaction\ k$$

$$n_k = number\ of\ transactions\ of\ class\ k$$

$$n_{ik} = number\ of\ transactions\ of\ class\ k\ at\ device\ i$$

$$m = number\ of\ classes$$

With the above definitions, the relative frequencies are defined as solutions to

$$\rho_{jk} = \sum_{i=1}^{N} \rho_{ik}\ p_{ijk} \qquad (6.155)$$

and

$$x_{ik} = \rho_{ik}\ \tau_{ik} \qquad (6.156)$$

The generating function for processing station i is given by

$$f_i(t_1, t_2, \cdots, t_m) = 1 + (x_{i1}t_1 + x_{i2}t_2 + \cdots + x_{im}t_m) +$$
$$+ (x_{i1}t_1 + x_{i2}t_2 + \cdots + x_{im}t_m)^2 + \cdots \qquad (6.157)$$

We can write Equation (6.157) as follows

$$f_i(t_1, t_2, \cdots, t_m) = \sum_{j=0}^{\infty} (x_{i1}t_1 + x_{i2}t_2 + \cdots + x_{im}t_m)^j \qquad (6.158)$$

And therefore, the generating function of the network is given by

$$g(t_1, t_2, \cdots, t_m) = \prod_{i=1}^{N} \sum_{j=0}^{\infty} \left(\sum_{k=1}^{m} x_{ik}\ t_k \right)^j \qquad (6.159)$$

or

$$
\begin{aligned}
g(t) = &[\, 1 + x_{11}t_1 + x_{12}t_1 + \cdots + x_{1m}t_1 + \\
&(x_{11}t_1 + x_{12}t_1 + \dots + x_{1m}t_1)^2 + \\
&(x_{11}t_1 + x_{12}t_1 + \cdots + x_{1m}t_1)^3 + \cdots \,] \times \\[6pt]
&[\, 1 + x_{21}t_2 + x_{22}t_2 + \dots + x_{2m}t_2 + \\
&(x_{21}t_2 + x_{22}t_2 + \dots + x_{2m}t_2)^2 + \\
&(x_{21}t_2 + x_{22}t_2 + \cdots + x_{2m}t_2)^3 + \cdots \,] \times \\[6pt]
&[\, 1 + x_{31}t_3 + x_{32}t_3 + \cdots + x_{3m}t_3 + \\
&(x_{31}t_3 + x_{32}t_3 + \cdots + x_{3m}t_3)^2 + \\
&(x_{31}t_3 + x_{32}t_3 + \cdots + x_{3m}t_3)^3 + \cdots \,] \times \\[6pt]
&\qquad\qquad\qquad \cdots \\[6pt]
&[\, 1 + x_{s1}t_m + x_{s2}t_m + \cdots + x_{sm}t_m + \\
&(x_{s1}t_m + x_{s2}t_m + \cdots + x_{sm}t_m)^2 + \\
&(x_{s1}t_m + x_{s2}t_m + \cdots + x_{sm}t_m)^3 + \cdots \,]
\end{aligned}
$$

$$(6.160)$$

As we saw for the cases with one and two classes of transactions, the coefficients of the generating function of the network play an important role in the calculation of performance indices. To arrive at a basic understanding of the generation of these coefficients, let us derive the iterative formula for this case.

5.1   Calculation of G Coefficients

Let us assume that we have in the network m different classes of transactions being processed. In this case the generating function of the network is given by Equation (6.160). Let us designate:

$$
g_1(t_1, t_2, \cdots, t_m) = f_1(t_1, t_2, \cdots, t_m)
$$

$$
g_2(t_1, t_2, \cdots, t_m) = g_1(t_1, t_2, \cdots, t_m)\, f_2(t_1, t_2, \cdots, t_m)
$$

and in general we have

$$g_i(t_1,t_2,\cdots,t_m) = g_{i-1}(t_1,t_2,\cdots,t_m) \, f_i(t_1,t_2,\cdots,t_m) \qquad (6.161)$$

We can write

$$f_i(t_1,t_2,\cdots,t_m) = \frac{1}{1 - x_{i1}t_1 - x_{i2}t_2 - \cdots - x_{im}t_m} \qquad (6.162)$$

Combining Equations (6.161) and (6.162), we obtain

$$\begin{aligned}
g_i(t_1,t_2,\cdots,t_m) = \; & g_{i-1}(t_1,t_2,\cdots,t_m) \\
& + \; x_{i1}\,t_i\,g_i(t_1,t_2,\cdots,t_m) \\
& + \; x_{i2}\,t_2\,g_i(t_1,t_2,\cdots,t_m) \\
& \quad\cdot\quad\cdot\quad\cdot\quad\cdot\quad\cdot\quad\cdot \\
& + \; x_{im}\,t_m\,g_i(t_1,t_2,\cdots,t_m) \qquad (6.163)
\end{aligned}$$

Let $G_i(n_1,n_2,\cdots,n_m)$ be the coefficient of $t_1^{n_1}t_2^{n_2}\cdots t_m^{n_m}$ in $g_i(t_1,t_2,\cdots,t_m)$; then,

$$\begin{aligned}
G_i(n_1,n_2,\cdots,n_m) = \; & G_{i-1}(n_1,n_2,\cdots,n_m) \\
& + \; x_{i1}\,G_i(n_1-1,n_2,\cdots,n_m) \\
& + \; x_{i2}\,G_i(n_1,n_2-1,\cdots,n_m) \\
& \quad\cdot\quad\cdot\quad\cdot\quad\cdot\quad\cdot\quad\cdot \\
& + \; x_{im}\,G_i(n_1,n_2,\cdots,n_m-1) \qquad (6.164)
\end{aligned}$$

Two important consequences can be drawn from Equation (6.168)

• All the $G_i(0,0,\ldots,0)$ are equal to one, that is,

$$G_i(0,0,\ldots,0) = 1 \qquad (i=1,2,3,\ldots,N)$$

• The coefficients $G_1(\ldots)$ are given by

$$G_1(j,0,\ldots,0) = x_{11}\,G_1(j-1,0,\ldots,0)$$
$$G_1(0,j,\ldots,0) = x_{12}\,G_1(0,j-1,\ldots,0)$$

$$\quad\cdot\quad\cdot\quad\cdot\quad\cdot\quad\cdot\quad\cdot\quad\cdot$$

$$G_1(0,0,\ldots,j) = x_{1m}\,G_1(0,0,\ldots,j-1)$$

since $G_0(\ldots) = 0$.

5.1.1 Processing Utilization

The utilization of the ith processing station is obtained by replacing

$f_i(t_1, t_2, \cdots, t_m)$ by $f_i(t_1, t_2, \cdots, t_m) - 1$ in the generating function of the network, Equation (6.159). Therefore,

$$h_i(t_1, t_2, \cdots, t_m) = \frac{g(t_1, t_2, \cdots, t_m)}{f_i(t_1, t_2, \cdots, t_m)} [f_i(t_1, t_2, \cdots, t_m) - 1] \quad (6.165)$$

With Equation (6.166), Equation (6.169) becomes

$$h_i(t_1, t_2, \cdots, t_m) = g(t_1, t_2, \cdots, t_m) \times$$
$$[x_{i1}t_1 + x_{i2}t_2 + \cdots + x_{im}t_m] \quad (6.166)$$

and the utilization, which is the coefficient of $t_1^{n_1} t_2^{n_2} \cdots t_m^{n_m}$ in $h_i(t_1, t_2, \cdots, t_m)$ divided by the coefficient of $t_1^{n_1} t_2^{n_2} \cdots t_m^{n_m}$ in $g(t_1, t_2, \cdots, t_m)$, is

$$U_i(n_1, n_2, \cdots, n_m) =$$
$$= [x_{i1} G(n_1 - 1, n_2, \cdots, n_m) + \quad (6.167)$$
$$x_{i2} G(n_1, n_2 - 1, \cdots, n_m) + \cdots +$$
$$x_{im} G(n_1, n_2, \cdots, n_m - 1)] / [G(n_1, n_2, \cdots, n_m)]$$

The above equation can be written as follows:

$$U_i(n_1, n_2, \ldots, n_m) =$$
$$= \frac{x_{i1} G(n_1 - 1, n_2, \cdots, n_m)}{G(n_1, n_2, \cdots, n_m)} + \frac{x_{i2} G(n_1, n_2 - 1, \cdots, n_m)}{G(n_1, n_2, \cdots, n_m)}$$
$$+ \cdots + \frac{x_{im} G(n_1, n_2, \cdots, n_m - 1)}{G(n_1, n_2, \cdots, n_m)}$$

The last equation allows us to state the following *lemmas*.

Lemma I - *The utilization of processor i in a closed network, holding a finite number of transactions of m different classes, is given by*

$$U_i(n_1, n_2, \cdots, n_m) = U_{i1}(n_1, n_2, \cdots, n_m) + U_{i2}(n_1, n_2, \cdots, n_m)$$
$$+ \cdots + U_{im}(n_1, n_2, \cdots, n_m)$$

Lemma II - *The determination of each individual utilization can be accomplished for each class, holding constant the total number of transactions n in the network.*

Lemma III - *The total number of transactions in the network is equal to* $\sum n_i$, *for* $i = 1, 2, 3, \ldots, m$, *or*

$$\sum_{i=1}^{m} n_i = n$$

The device utilization for m classes is designated as $U_i(n_1,n_2,....,n_m)$, where i stands for the device and m for the number of classes in the processing network. We know that for an analytical model the following condition must be met

$$\lim_{n \to \infty} U_i(n_1,n_2,....,n_m) = 1.0$$

where $n = \sum_{i=1}^{m} n_i$.

Since we cannot have an infinite number of transactions in the processing network, nor have the network saturated, some practical limitations must to be imposed on the maximum acceptable utilization; let us say for practical purposes, 0.96. This is a value that has been used extensively for the modeling of local computer networks.

For a closed network, the number of transactions which brings $U_i(n)$ to a value of at least 0.96 will be the **capacity** of the processing network.

Similarly, to obtain the overlap between processing stations i and j, we have

$$h_{ij}(t_1,t_2, \cdots ,t_m) =$$

$$= \frac{g(t_1,t_2, \cdots ,t_m)}{f_i(t_1,t_2, \cdots ,t_m) \, f_j(t_1,t_2, \cdots ,t_m)} \times$$

$$[f_i(t_1,t_2, \cdots ,t_m) - 1] \, [f_j(t_1,t_2, \cdots ,t_m) - 1] \qquad (6.168)$$

then

$$U_{ij}(n_1,n_2, \cdots ,n_m) = \frac{\displaystyle\sum_{nu=1}^{m-1} \sum_{k=\nu+1}^{m} [x_{i\nu}x_{jk} + x_{ik}x_{j\nu}] \, t_\nu + k}{G(n_1,n_2, \cdots ,n_m)} \qquad (6.169)$$

The generating function of the processing network, for the case of N products, sharing the processing stations is given by:

$$g(t_1,t_2,t \ldots ,t_N) =$$

$$\prod_{j=1}^{s} \frac{1}{1 - x_{1j}t_1 - x_{2j}t_2 - x_{3j}t_3 - \cdots - x_{Nj}t_N} \qquad (6.170)$$

The partial fraction expansion of Equation (6.170) is straight forward, following the procedure presented for $m=2$.

Equation (6.160) can be written as follows:

$$\prod_{j=1}^{s} \frac{1}{1-(x_{j1}t_1 + x_{j2}t_2 + \cdots x_{jm}t_m)} =$$

$$\sum_{n_1}^{\infty} \cdots \sum_{n_m}^{\infty} G(n_1,n_2,\cdots,n_m)\; t_1^{n_1} t_2^{n_2} \cdots t_m^{n_m} \qquad (6.171)$$

Assuming that all x_{ji} for $j=1,2,...,s$ and $i=1,2,...,m$ are different, we can establish the following result

$$G(n_1,n_2,....,n_m) =$$

$$\sum_{a_1^1=0}^{n_1} \sum_{a_2^1=0}^{n_1-a_1^1} \cdots \sum_{a_{s-1}^1=0}^{n_1-\sum_{j=1}^{s-1}a_j^1} \cdots \sum_{a_1^{m-1}=0}^{n_{m-1}} \sum_{a_2^{m-1}=0}^{n_{m-1}-a_1^{m-1}} \cdots \sum_{a_{s-1}=0}^{n_{n-1}-\sum_{j=1}^{s-2}a_j^{n-1}} \sum_{k_1=\sum_{i=1}^{n-1}a_1^i}^{\sum_{i=1}^{m}n_i} \sum_{k_2=\sum_{i=1}^{m-1}a_2^i}^{\sum_{i=1}^{m}n_i-k_1} \cdots \sum_{k_{s-1}=\sum_{i=1}^{m-1}a_{s-1}^i}^{\sum_{i=1}^{m-1}n_i-\sum_{j=1}^{s-2}k_j}$$

$$\times \left[\prod_{j=1}^{s-1} \begin{bmatrix} k_j \\ a_j^1,a_j^2,\cdots,a_j^{m-1} \end{bmatrix} (x_{j1})^{a_j^1} \cdots (x_j(m-1))^{a_j^{m-1}} (x_{jm})^{k_j-\sum_{i=1}^{m-1}a_j^i} \right]$$

$$\times \left[\begin{matrix} \sum_{i=1}^{m}n_i-\sum_{j=1}^{s-1}k_j \\ (n_1-\sum_{j=1}^{s-1}a_j^1),\cdots,(n_{m-1}-\sum_{j=1}^{s-1}a_j^{m-1}) \end{matrix} \right] (x_{s1})^{n_1-\sum_{j=1}^{s-1}a_j^1} \cdots \times$$

$$[x_{s(m-1)}]^{n_1-\sum_{j=1}^{s-1}a_j^{m-1}} (x_{sm})^{n_m-\sum_{j=1}^{s-1}(k_j\sum_{i=1}^{m-1}a_j^i)}$$

where the multilinear coefficient is denoted as

$$\begin{bmatrix} k \\ a^1,a^2,\cdots,a^{m-1} \end{bmatrix} = \frac{k!}{a^1!\,a^2!\,\cdots a^{m-1}!\,(k-\sum_{i=1}^{m-1}a_i)!} \qquad (6.173)$$

To derive this formula we first expand each factor in the product into its corresponding geometric series

$$\prod_{j=1}^{s} \frac{1}{1 - (x_{j1}t_1 + x_{j2}t_2 + \cdots + x_{jm}t_m)} =$$
$$\prod_{j=1}^{s} \sum_{k_j=0}^{\infty} (x_{j1}t_1 + x_{j2}t_2 + \cdots + x_{jm}t_m)^{k_j} \qquad (6.174)$$

Each of the above powers is expanded by the multinomial theorem

$$(x_{j1}t_1 + x_{j2}t_2 + \ldots + x_{jm}t_m)^{k_j} =$$

$$\sum_{k_j=0}^{\infty} \sum_{a_j^1=0}^{k_j} \cdots \sum_{a_j^{m-1}=0}^{k_j - \sum_{i=1}^{m-2} a_j^i} \left[\begin{matrix} k_j \\ a_j^1, a_j^2, \cdots, a_j^{m-1} \end{matrix} \right] (x_{j1}t_1)^{a_j^1} \cdots \times$$

$$[x_{j(m-1)}t_{m-1}]^{a_j^{m-1}} (x_{jm}t_m)^{k_j - \sum_{i=1}^{m-1} a_j^i}$$

When the sums replace the powers, the resulting series can be rearranged to yield

$$\prod_{j=1}^{s} \frac{1}{1 - (x_{j1}t_1 + x_{j2}t_2 + \cdots + x_{jm}t_m)} =$$

$$\sum_{k_1=0}^{\infty} \sum_{a_1^1=0}^{\infty} \cdots \sum_{a_j^{m-1}}^{k_1 - \sum_{i=1}^{m-2} a_1^i} \cdots \sum_{k_s=0}^{\infty} \sum_{a_s^1=0}^{k_s} \cdots \sum_{a_s^{m-1}=0}^{k_s - \sum_{i=1}^{m-2} a_s^i}$$

$$\left[\prod_{j=1}^{s} \left[\begin{matrix} k_j \\ a_j^1, a_j^2, \cdots, a_j^{m-1} \end{matrix} \right] (x_{j1})^{a_j^1} \cdots [x_{j(m-1)}]^{a_j^{m-1}} (x_{jm})^{k_j - \sum_{i=1}^{m-1} a_j^i} \right]$$

$$t_1^{\sum_{j=1}^{s} a_j^1} t_2^{\sum_{j=1}^{s} a_j^2} \cdots t_{m-1}^{\sum_{j=1}^{s} a_j^{m-1}} t_m^{\sum_{j=1}^{s} k_j - \sum_{j=1}^{s} \sum_{i=1}^{m-1} a_j^i}$$

Notice that all powers of t_i are collected and the exponents are added. We now let

$$n_i a = \sum_{j=1}^{s} a_j^i \qquad \text{for } i = 1, 2, \ldots, m-1 \qquad \text{and} \qquad (6.177)$$

$$n_m = \sum_{j=1}^{s} k_j - \sum_{j=1}^{s} \sum_{i=1}^{m-1} a_j^i = \sum_{j=1}^{s} k_j - \sum_{i=1}^{m-1} m_i \qquad (6.178)$$

to represent the series in the form needed in Equation (6.176). To introduce the n's into the summation we must eliminate some of the other variables of summation as follows

$$a_s^i = n_i - \sum_{j=1}^{s-1} a_j^i \qquad \text{for } i = 1, 2, \ldots, m-1 \qquad (6.179)$$

$$k_s = \sum_{i=1}^{m} n_i - \sum_{j=1}^{s-1} k_j \qquad (6.180)$$

Before making this change of variables in Equation (6.176), we perform a preliminary exchange of the order of summation for convenience. Notice that the summations in Equation (6.176) consist of s groups of n summations of the form

$$\sum_{k=0}^{\infty} \sum_{a^1=0}^{k} \sum_{a^2=0}^{k} \cdots \sum_{a^{m-1}=0}^{k-\sum_{i=1}^{m-2} a^i} \qquad (6.181)$$

It would be convenient if the k-indexed sum occurred last; in order to accomplish this we invoke the following lemma.

Lemma IV -

$$\sum_{k=A}^{\infty} \sum_{a=0}^{k-A} [\cdots] = \sum_{a=0}^{\infty} \sum_{k=a+A}^{\infty} [\cdots] \qquad (6.182)$$

Proof - The first sum represents the set $\{ (k,a) : A \leqslant k < \infty$ and for each k, $0 \leqslant a \leqslant k - A \}$. Note that $k \geqslant a + A$ and $k \geqslant A$ if and only if $a \geqslant 0$, so this set can be written as $\{ (k,a) : 0 \leqslant a \leqslant \infty$ and for each a, $a + A \leqslant k \leqslant \infty \}$. This set is represented by the second sum and the equality is established.

If *Lemma 1* is applied $(m-1)$ times in each group and then the change of variables in Equations (6.154) and (6.155) is made in Equation (6.110), the result is

$$\prod_{j=1}^{s}\frac{1}{1-(x_{j1}t_1+x_{j2}t_2+\cdots+x_{jm}t_m)}=$$

$$\sum_{a_1^1=0}^{\infty}\sum_{a_1^2=0}^{\infty}\cdots\sum_{a_1^{m-1}=0}^{\infty}\sum_{k_1=\sum_{i=1}^{m-1}a_1^i}^{\infty}\cdots\sum_{a_{s-1}^1=0}^{\infty}\cdots\sum_{k_{s-1}=\sum_{i=1}^{m-1}a_{s-1}^i}^{\infty}$$

$$\sum_{n_1-\sum_{j=1}^{s-1}a_j^1=0}^{\infty}\sum_{n_2-\sum_{j=1}^{s-1}a_j^2=0}^{\infty}\cdots\sum_{n_{m-1}-\sum_{j=1}^{s-1}a_j^{m-1}=0}^{\infty}\sum_{n_m+\sum_{i=1}^{m-1}n_i-\sum_{j=1}^{s-1}k_j=\sum_{i=1}^{n-1}[n_i-\sum_{j=1}^{s-1}a_j^i]}^{\infty}$$

$$\left[\prod_{j=1}^{s-1}\binom{k_j}{a_j^1,a_j^2,\cdots,a_j^{m-1}}(x_{j1})^{a_1}\cdots(x_{jm})^{k_j-\sum_{i=1}^{m-1}a_j^i}\right]$$

$$\left[\left[\binom{\sum_{i=1}^{m}n_i-\sum_{j=1}^{s-1}k_j}{(n_i-\sum_{j=1}^{s-1}a_j^i)\cdots(n_{m-1}-\sum_{j=1}^{s-1}a_j^{m-1})}\right](x_{s1})^{n_1-\sum_{j=1}^{s-1}a_j^1}\cdots(x_{jm})^{n_m-\sum_{j=1}^{s-1}(k_j-\sum_{i=1}^{m-1}a_j^i)}\right]$$

$$t_1^{n_1}t_2^{n_2}\cdots t_m^{n_m} \tag{6.183}$$

Notice that because of the change of variables, the terms corresponding to (x_{si}) had to be written separately. Moreover the power of (x_{sm}) is simplified

$$\sum_{i=1}^{m}m_i-\sum_{j=1}^{s-1}k_j-\sum_{i=1}^{m-1}[m_i-\sum_{j=1}^{s-1}a_j^i]=$$

$$m_m-\sum_{j=1}^{s-1}(k_j-\sum_{i=1}^{m-1}a_j^i) \tag{6.184}$$

as can be the final summation involving n_m.

The expression in Equation (6.183) is nearly of the form needed to identify the coefficients $G(n_1,n_2,....,n_m)$ in Equation (6.160), except that the indices $n_1,n_2,....,n_m$ occur in the wrong location in the sums. In order to shift them to the front of the summation we will interchange some limits of summation. The form of the summand is exactly as stated in Equation (6.183) so it is only the order of summation that must be accounted for. To this end we work only with the summations beginning with that of Equation (6.183)

$$\sum_{a_1^1=0}^{\infty} \sum_{a_1^{m-1}=0}^{\infty} \sum_{k_1=\sum_{i=1}^{m-1} a_1^i} \cdots \sum_{a_{s-1}^1=0}^{\infty} \cdots \sum_{a_{s-1}^{m-1}=0}^{\infty} \sum_{k_{s-1}=\sum_{i=1}^{m-1} a_{s-1}^i}$$

$$\sum_{k_1=\sum_{i=1}^{m-1} a_1^i} \cdots \sum_{k_{s-1}=\sum_{i=1}^{m-1} a_{s-1}^i} \sum_{n_m=\sum_{j=1}^{s-1}(k_j-\sum_{i=1}^{m-1} a_j^i)} \tag{6.185}$$

We first regroup terms so that each sum indexed by n_i, for $i=1,2,...,m-1$, is the last in a group of s sums containing the $a_1^i, a_2^i, \cdots, a_{s-1}^i$ indexed sums. Similarly the n_m indexed sum follows those sums indexed by $k_1, k_2, \cdots, k_{m-1}$. These changes are allowed because they preserve the rule that all indices of summation must precede their occurrence in a limit of summation. The result of this regrouping in Equation (6.173) yields

$$\sum_{a_1^1=0}^{\infty} \sum_{a_2^1=0}^{\infty} \cdots \sum_{a_{s-1}^1=0}^{\infty} \sum_{n_1=\sum_{j-1}^{s-1} a_j^1} \cdots \sum_{a_1^{m-1}=0}^{\infty} \cdots \sum_{a_{s-1}^{m-1}=0}^{\infty} \sum_{n_{m-1}=\sum_{j=1}^{s-1} a_j^{m-1}}$$

$$\sum_{k_1=\sum_{i=1}^{m-1} a_1^i} \cdots \sum_{k_{s-1}=\sum_{i=1}^{m-1} a_{s-1}^i} \sum_{n_m=\sum_{j=1}^{s-1}(k_j-\sum_{i=1}^{m-1} a_j^i)} \tag{6.186}$$

Lemma 4 (with k replaced by n) is applied $(s-1)$ times in each of the first $(m-1)$ groups of sums to move the n_i indices to the left for $i=1,2,...,m-1$. To move the n_m index to the left of the k's we apply the following Lemma $(s-1)$ times.
Lemma 5 -

$$\sum_{k=A}^{\infty} \sum_{n=(k-A)+B}^{\infty} = \sum_{n=B}^{\infty} \sum_{k=A}^{(n-B)+A} \tag{6.187}$$

The proof is similar to that of *Lemma 4*. When these two lemmas are applied to Equation (6.183), the result is

$$\sum_{n_1=0}^{\infty} \sum_{a_1^1=0}^{n_1} \cdots \sum_{a_{s-1}^1=0}^{n_1-\sum_{i=1}^{m-2}a_i^1} \cdots \sum_{n_{m-1}=0}^{\infty} \sum_{a_1^{m-1}=0}^{n_{m-1}} \cdots$$

$$\sum_{a_{s-1}^{m-1}=0}^{n_{m-1}-\sum_{i=1}^{m-2}a_i^{m-1}} \sum_{n_m=0}^{\infty} \sum_{k_1=\sum_{i=1}^{m-1}a_1^i}^{\sum_{i=1}^{m}n_i} \cdots \sum_{k_{s-1}=\sum_{i=1}^{m-1}a_{s-1}^i}^{\sum_{i=1}^{m}n_i-\sum_{j=1}^{s-2}k_j} \tag{6.188}$$

The final rearrangement involves moving all the n indexed sums to the front as is needed for Equation (6.107). Reinsertion of the summand in Equation (6.105) into the series with the order of summation as in Equation (6.120) yields the series from which the coefficients $G(n_1,n_2,....,n_m)$ as described in Equation (6.118) are obtained.

Example 6.5 - Consider the local computer network shown in Fig. 6.6, with 4 different classes of transactions, and the service times shown in Table 6.10. Under these conditions we obtain the utilizations, shown in Table 6.11, for the indicated network state.

Linear Approximation - The coefficients of the linear regression are given by a = 13.32268   b = 4.37800 then we have the values shown in Table 6.12. For the linear regression coefficients a = 14.19599 b = 4.31300 the corresponding values are given in Table 6.13. If the linear regression coefficients are: a = 3.16643 b = 1.30643 then Table 6.14 contains the linear model errors.

Table 6.10   Service Times per Transaction Class

Device	Class 1	Class 2	Class 3	Class 4
CPU	3.20	2.98	4.17	1.66
I/O1	3.85	5.23	4.17	2.66
I/O2	4.21	3.33	4.27	2.23
I/O3	3.26	4.16	4.11	1.98
I/O4	3.72	3.79	3.99	2.52

Table 6.11 Utilizations Results

State	Class	CPU	I/O1	I/O2	I/O3	I/O4
(8,3,2,4)	1	47.38	13.81	13.79	13.80	13.79
	2	17.40	7.39	4.30	6.94	5.54
	3	17.04	2.91	2.72	3.39	2.88
	4	23.17	8.99	6.88	7.90	8.81
(7,3,2,4)	1	44.09	12.85	12.83	12.84	12.84
	2	18.47	7.85	4.56	7.37	5.88
	3	12.81	3.10	2.90	3.61	3.07
	4	24.60	9.55	7.31	8.39	9.35
(6,3,2,4)	1	40.36	11.76	11.75	11.75	11.75
	2	19.69	8.37	4.87	7.86	6.27
	3	13.70	3.32	3.10	3.86	3.28
	4	26.23	10.18	7.79	8.94	9.97
(5,3,2,4)	1	36.08	10.52	10.50	10.51	10.50
	2	21.09	8.97	5.21	8.41	6.72
	3	14.72	3.56	3.33	4.15	3.53
	4	28.08	10.90	8.35	9.57	10.68
(4,3,2,4)	1	31.14	9.07	9.06	9.07	9.06
	2	22.70	9.65	5.61	9.06	7.23
	3	15.91	3.85	3.60	4.48	3.81
	4	30.22	11.73	8.98	10.30	11.49
(3,3,2,4)	1	25.36	7.39	7.38	7.38	7.38
	2	24.58	10.45	6.07	9.81	7.83
	3	17.32	4.19	3.92	4.88	4.15
	4	32.72	12.70	9.72	11.16	12.44
(2,3,2,4)	1	18.49	5.39	5.38	5.38	5.38
	2	26.79	11.39	6.62	10.69	8.53
	3	18.99	4.60	4.30	5.35	4.55
	4	35.67	13.85	10.60	12.16	13.56

Table 6.12 Errors Using Linear Models

n	$U(n)$	$U_c(n)$	$U_c(n) - U(n)$	$[U_c(n) - U(n)]/U(n)$
3	25.360	26.457	1.097	4.324
4	31.140	30.835	−.305	−.981
5	36.080	35.213	−.067	−.240
6	40.360	39.591	−.769	−.196
7	44.090	43.969	−.121	−.275
8	47.380	48.347	0.967	2.040

Table 6.13 Errors Using Linear Models

n	$U(n)$	$U_c(n)$	$U_c(n) - U(n)$	$[U_c(n) - U(n)]/U(n)$
4	31.140	31.448	0.308	0.889
5	36.080	35.761	$-.319$	$-.884$
6	40.360	40.074	$-.286$	$-.709$
7	44.090	44.387	0.297	0.674

Table 6.14 Errors Using Linear Models

n	$U(n)$	$U_c(n)$	$U_c(n) - U(n)$	$[U_c(n) - U(n)]/U(n)$
2	5.380	5.939	0.559	10.396
3.	7.380	7.326	-.054	-.730
4	9.060	8.712	-.348	-3.839
5	10.500	10.099	-.401	-3.823
6	11.750	11.485	-.265	-2.255
7	12.830	12.871	0.041	0.383
8	13.790	14.258	0.468	3.393

5.1.2 Latest Developments in Analyzing Queuing Networks

There are some recent approaches for analyzing queuing networks. In general these more recent methods fall outside the scope of this book which is concerned with the development of analytical queuing models that can be implemented with minimum effort, both in terms of knowledge and input data requirements. The latest developments fall into three areas. First, there are **Approximate MVA methods for large closed queuing networks** which present a simple approach for the determination of network performance [10,35]. A discussion of this algorithm is presented below. The second area is **Asymptotic Expansions of Queuing Networks**. The reader is referred to [32,33] for a discussion of asymptotic expansions and to [34] for a software package implementation. The third area, **Decomposition Method**, consists of an analysis based on the method of decomposition, in which the total network is broken up into subsystems. A complete discussion on this subject can be found in [25].

Although all these techniques are beyond our current scope, we will discuss the MVA method in some detail because it can be used in conjunction with Buzen's approach.

6. MEAN VALUE ANALYSIS APPROACH

In this section we will discuss the single iteration mean value analysis algorithm (SI-MVA) presented by K. Mani Chandy and P. Neuse [10]. This algorithm presents a simple approach for network performance evaluation, but requires more data for network description than Buzen's model. We will first discuss the mathematical background, trying to maintain the nomenclature already in use. We then discuss the iteration procedure, which will be followed with a computer network application.

Before starting our discussion, we introduce the concept of a *delay queue*. This queue provides service immediately (without delay) to all transactions that reach it. This particular queue does not have contention of any kind. We describe this queue as one having an infinite number of servers. Delay queues are used to model terminals, since there is normally one terminal per transaction.

Associated with each server, we have a queue as illustrated in Fig. 6.16. When a transaction of some class k reaches this queue, it waits for service until the server becomes free for processing. At that time a new transaction from the queue moves into the server. During the discussion on Buzen's model, we assumed that the transaction receiving service was considered part of the queue.

In general we can state the methodology for modeling computer devices in a processing network as follows:

- **Single-Server Queue** - This type of queue is used to model a *fixed-rate processor*.
- **Delay Queue** - This type of queue is used to model terminals.

It has been shown that the *product-form solution* holds for networks in equilibrium consisting of single-server queues or delay queues. This type of networks is called *basic networks*.

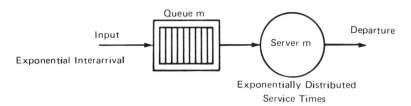

Figure 6.16   Processing Station *m*.

Before starting the mathematical background, we introduce some definitions:

n = total number the transactions in the network

k = number of transaction classes

n_j = number of transactions of class j in the network

ν_{mk} = visit ratio of class k transactions at queue m

m = number of processing stations in the network

τ_{mk} = mean service time of class k transactions at processing station m

$\bar{e}_k$ = unit vector with a 1 in the kth place and zeros elsewhere

w_{mk} = mean waiting time for transactions of class k at the queue of processing station m

Q_k = type of queue k: single server or delay

y_{mk} = mean number of transactions of class k that pass through queue m per unit of time

$L_{(m)}$ = the mean number of transactions in queue at processing station m, regardless of transaction classes

$L_{mk}(n)$ = the mean number of transactions of class k in queue at processing station m

The total number of transactions in the network is given by

$$n = n_1 + n_2 + \cdots + n_k = \sum_{j=1}^{k} n_j \qquad (6.189)$$

Also, let $\bar{n}$ be the vector whose components are $(n_1, n_2, n_3, \cdots, n_k)$, or

$$\bar{n} = (n_1, n_2, n_3, \cdots, n_k) \qquad (6.190)$$

The mean number of transactions in queue at the processing station m is given by

$$L_{(m)}(n) = L_{m1}(n) + L_{m2}(n) + \cdots + L_{mk}(n) \qquad (6.191)$$

The average time spent by a transaction of class k in the combination queue-server is given by

$$w_{mk}(n) = \tau_{mk} + \tau_{waiting} \qquad (6.192)$$

where $\tau_{waiting}$ is the average time spent by a class k transaction in the queue until it is ready to move to the server. This waiting time can be written as

$$\tau_{waiting} = L_{(m)} \, \tau_{mk} \, (\bar{n} - \bar{e}_k) \qquad (6.193)$$

Replacing Equation (6.193) into Equation (6.192), we obtain

$$w_{mk}(n) = \tau_{mk} \, (1 + L_{(m)} \, (\bar{n} - \bar{e}_k)) \qquad (6.194)$$

Reiser and Lavenberg [35] showed that the equilibrium performance statistics for a basic network are given by Equation (6.194). Mani Chandy and Neuse mentioned in their paper [10] *that Equation (6.194) has intuitive appeal.*

Let us briefly discuss the preceding equations:

- The number of transactions sharing the server at the queue of processing station m, with a specific class k transaction, receiving service is

$$L_{(m)} \, (\bar{n} - \bar{e}_k) \qquad (6.195)$$

which is the mean number of transactions in the queue of processing station m when the class k transaction is removed from the population.

- On the average, class k transactions receive service for only a fraction of time, designated as ξ_f, and defined as follows

$$\xi_f = \frac{\tau_{mk}}{w_{mk}(n)} = \frac{1}{1 + L_{(m)} \, (\bar{n} - e_{\bar{k}})} \qquad (6.196)$$

- It is evident that for a delay queue (infinite server), Equation (6.194) becomes

$$w_{mk}(n) = \tau_{mk} \qquad (6.197)$$

According to Little's law *the average number of transactions in the system is equal to the average arrival rate of transactions to that system times the average time spent in that system.* It is intuitive that an arriving transaction should find the same average number $\bar{N}$ in the system as it leaves behind upon its departure. This latter quantity is simply the arrival rate times the average time in system T. Then,

$$\bar{N} = \lambda \, T$$

Taking into consideration Little's law, we can write

$$L_{mk}(n) = y_{mk}(n) \, w_{mk}(n) \qquad (6.198)$$

Since we consider closed networks, each particular class of transaction comes to the same queue over and over again. Let $c_{mk}(n)$ denote the queue m class k cycle time, i.e., the average time between successive arrivals of the same class k transaction at queue m. For each trip to queue m, a class k transaction visits queue m' an average of $\nu_{m'k}/\nu_{mk}$ times and so spends an average of $w_{m'k}(n) \, \nu_{m'k}/\nu_{mk}$ time units in queue m'. Thus, the time required

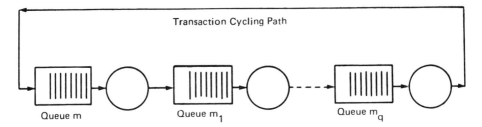

Figure 6.17   Transaction Cycling Path.

for a class k transaction to reach queue m, after it leaves the associated server of queue m, is given by

$$c'_{mk}(n) = \sum_{m'=1}^{q} w_{m'k}\, \nu_{m'k}/\nu_{mk} \qquad (6.199)$$

where q is the number of queues in the path of class k transactions. The summation over m' extends to all the queues involved in the transaction cycling path, as shown in Fig. 6.17. Therefore, the total cycle time per class k transaction, for device m, is given by

$$c_{mk}(n) = c'_{mk}(n) + w_{mk}(n) \qquad (6.200)$$

Little's law applied to class k over the entire network, as viewed from queue m, yields

$$n_k = y_{mk}(n)\, c_{mk}(n) \qquad (6.201)$$

Taking into consideration the previous equations, we can write

Table 6.15   SI-MVA

$$w_{mk}(n) = \tau_{mk}\,[1 + L_{(m)}(\bar{n} - e_k)]$$

$$w_{mk}(n) = \tau_{mk}$$

$$L_{mk}(n) = y_{mk}(n)\, w_{mk}(n)$$

$$L_{mk}(n) = n_k\, \frac{\nu_{mk}\, w_{mk}(n)}{w_{mk}(n) + \sum\limits_{m'=}^{q} \nu_{m'k}\, w_{m'k}(n)}$$

$$y_{mk}(n) = \frac{L_{mk}(n)}{w_{mk}(n)}$$

$$U_{mk}(n) = y_{mk}(n)\, \tau_{mk}$$

$$L_{mk}(n) = \frac{n_k\, w_{mk}(n)}{c_{mk}(n)}$$

$$= n_k \frac{\nu_{mk}\, w_{mk}(n)}{w_{mk}(n) + \sum_{m'=1}^{q} \nu_{m'k}\, w_{m'k}(n)} \qquad (6.202)$$

The class k utilization of queue m may be obtained from

$$U_{mk}(n) = y_{mk}(n)\, \tau_{mk} \qquad (6.203)$$

The equations associated with the Single Iteration Mean Value Analysis Algorithm (SI-MVA) are displayed in Table 6.15.

6.1 Single Iteration Mean Value Analysis

To use this technique we need the following information:

1. The number of transaction classes K and the number of service stations M.
2. The number of transactions for each class, n_k.
3. The value of Q_m for each queue m.
4. For each value of m and k, we need to know ν_{mk}, τ_{mk}, and $L_{(m)}\,(n - e_k)$

The computation procedure is as follows:

- **Step 1** - Choose a value for m and k.
- **Step 2** - Compute $W_{mk}(n)$:

$$w_{mk}(n) = \tau_{mk}\,[1 + L_{(m)}\,(n - e_k)] \qquad (6.204)$$

- **Step 3** - Compute $L_{mk}(n)$:

$$L_{mk}(n) = n_k \frac{\nu_{mk}\, w_{mk}(n)}{w_{mk} + \sum_{m'} \nu_{m'k}\, w_{m'k}(n)} \qquad (6.205)$$

- **Step 4** - Compute $y_{mk}(n)$:

$$y_{mk}(n) = \frac{L_{mk}(n)}{w_{mk}(n)} \qquad (6.206)$$

- **Step 5** - Compute $U_{mk}(n)$:

$$U_{mk}(n) = y_{mk}(n)\, \tau_{mk} \qquad (6.207)$$

- **Step 6** - Change the values of k and m and go back to Step 2.

Fig. 6.18 shows the single iteration mean-value process already described. The SI-MVA algorithm assumes that the statistics of populations $\bar{n} - \bar{e}_k$ are known.

Example 6.6 - Consider the basic computer network shown in Fig. 6.19. This computer network handles two different transaction classes: Class 1-TSO and Class 2-Batch. The flow of these transactions through the

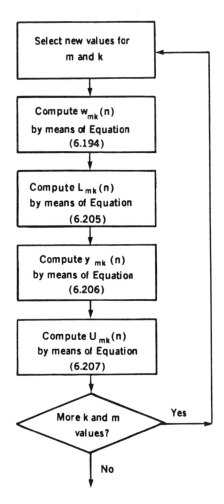

Figure 6.18   Mean-Value Analysis Process.

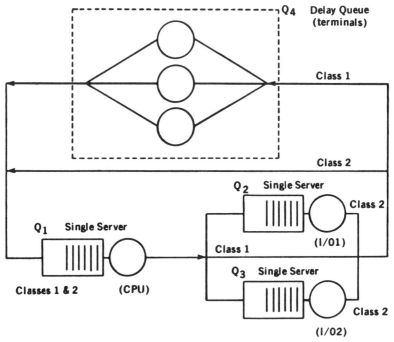

Figure 6.19 Information System.

network is defined. The devices in the network are 1 CPU (Single Server); 2 I/O (Single Server); and 1 Delay Queue (Infinite Server Terminals). This is a *closed network*, holding 5 class 1 transactions and 16 class 2 transactions.

The Q's in our basic computer network are defined as
Q_1 =   Single Server
Q_2 =   Single Server
Q_3 =   Single Server
Q_4 =   Delay Queue

The values of v_{ij} and s_{ij} are:

$$v_{11} = 1 \qquad v_{12} = 1$$
$$v_{21} = 0 \qquad v_{22} = 1$$
$$v_{31} = 0 \qquad v_{32} = 1$$

$$v_{41} = 1 \qquad v_{42} = 0$$
$$\tau_{11} = 3 \qquad \tau_{12} = 4$$
$$\tau_{21} = 0 \qquad \tau_{22} = 5$$
$$\tau_{31} = 0 \qquad \tau_{32} = 6$$
$$\tau_{41} = 6 \qquad \tau_{42} = 0$$

The number of transactions in the closed network is given by

$$n_1 = 5 \qquad n_2 = 16$$

The average queue length in each device is

$$L_{(1)} = 18.31 \qquad L_{(2)} = 0.10 \qquad L_{(3)} = 0.07 \qquad L_{(4)} = 0.0$$

The values for the performance indices appear in Table 6.16. We note that if we add the two values of $L_{1k}(n)$, we obtain

$$L'_{(1)} = L_{11} + L_{12} = 4.75 + 15.21 = 19.96$$

which is different from the value $L_{(1)}$. The difference occurs because $L'_{(1)}$ counts the transaction that is receiving service.

Table 6.16 Performance Indices

Device	Class	$w_{mk}(n)$ (sec)	$L_{mk}(n)$	$y_{mk}(n)$ (Trans./hr)	$U_{mk}(n)$ (%)
1	1	57.93	4.75	709.18	78.4
	2	77.24	15.21	295.18	24.5
2	1	0.00	0.00	0.00	0.0
	2	5.50	0.59	166.04	58.3
3	1	0.00	0.00	0.00	0.0
	2	6.42	0.53	129.14	49.0
4	1	6.00	0.10	141.82	18.1
	2	0.00	0.00	0.00	0.0

6.2   The Core Algorithm

The problem with the SI-MVA algorithm is that it requires a complete solution, i.e., solutions for all populations from $(0,0,0,....,0)$ up to $(n_1,n_2,n_3,.....,n_K)$. An alternative approach would be to use information about population n to *guess* statistics for population $n - e_j$ and then use the SI-MVA algorithm to compute new statistics for population n.

$$\bar{n} \rightarrow \bar{n} - \bar{e}_j \rightarrow \bar{n}$$

To estimate population $n - e_j$ statistics, the *Bard-Schweitzer Algorithm* assumes that $L_{mk}(n - e_j)$ is proportional to $L_{mk}(n)$. Then we can write

$$L_{mk}(\bar{n}-\bar{e}_j) = \begin{cases} L_{mk}(\bar{n}) & \text{for } k \neq j \\ \dfrac{n_j-1}{n_j} L_{mj}(\bar{n}) & \text{for } k=j \end{cases} \qquad (6.208)$$

In the heuristics for estimating population $n - e_k$ statistics from population n statistics, Chandy and Neuse define $F_{mk}(n)$ as the fraction of class k transactions at queue m for population n. Thus,

$$F_{mk}(\bar{n}) = \frac{L_{mk}(\bar{n})}{n_k} \qquad (6.209)$$

Then further define $D_{mkj}(\bar{n})$ to be the change in the fraction of class k transactions at queue m resulting from the removal of one class j transaction.

$$D_{mkj}(\bar{n}) = F_{mk}(\bar{n}-\bar{e}_j) - F_{mk}(\bar{n}) \qquad (6.210)$$

From Equations (6.207) and (6.208), the mean number of class k transactions at queue m given population $\bar{n} - \bar{e}_j$ can be expressed as

$$L_{mk}(\bar{n} - \bar{e}_j) = (\bar{n} - \bar{e}_j)_k [F_{mk}(\bar{n}) - D_{mkj}(\bar{n})] \qquad (6.211)$$

where $(\bar{n} - \bar{e}_j)_k$ is the population of class k when one class j transaction is removed from population n.

We cannot compute $D_{mkj}(\bar{n})$ using Equation (6.209) because it requires the unknown $L_{mk}(\bar{n}-\bar{e}_j)$ values. Instead, we estimate the $D_{mkj}(\bar{n})$ values and then compute approximations to the $L_{mk}(\bar{n}-\bar{e}_j)$'s using Equation (6.209).

The core algorithm appears in Fig. 6.20. The calculation process is as follows:

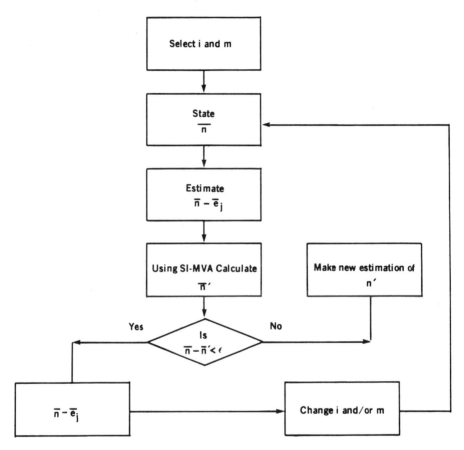

Figure 6.20   Core Algorithm.

1. **Initialization** - Set $I = 1$.
2. **Step 1** - Compute approximations to the $L_{mk}(\bar{n} - \bar{e}_j)$ from Equations (6.207) and (6.209) for all m, k, j.
3. **Step 2** - SI-MVA computes new estimates $L_{mk}(\bar{n})$ from $L_{mk}(\bar{n} - \bar{e}_j)$ using Equations (6.194), (6.197), and (6.203). $w_{mk}(\bar{n})$ estimates are also produced by these computations.
4. **Termination Test** - If

$$\max_{all\ m,k} \frac{|L_{mk}^{I} - L_{mk}^{I-1}|}{n_k} > \frac{1}{(4000 + 16|\bar{n}|)} \tag{6.212}$$

where the superscript I designates iteration I statistics; then set $I=I+1$ and go to **Step 1**. Otherwise, go to the **Final Step**.

5. **Final Step** - Compute throughput estimates $Y_{mk}(n)$ from Equation (6.198) and utilization estimates $U_{mk}(n)$ from Equation (6.205) for all m, k.

The Bard-Schweitzer proportionality assumption (6.187) is equivalent to assuming that $D_{mkj}(n) = 0$. Chandy and Neuse refer to the Core algorithm with inputs $D_{mkj}(n) = 0$ as the Schweitzer-Core algorithm.

6.3   The Linearizer Algorithm

The Linearizer algorithm is an improvement over the core algorithm, leading to more accurate estimates of $D_{mkj}(\bar{n})$. It invokes the Core algorithm as a subroutine and computes its own $D_{mkj}(\bar{n})$ estimates. It also assumes that $D_{mkj}(\bar{n}-\bar{e}_j) = D_{mkj}(\bar{n})$, which implies that F is a linear function of the population, hence, the name *Linearizer*. For a good description on this approach, see [10].

7.   FURTHER READING

An excellent work on queuing theory is the book by Kleinrock [23]. There is also a lot of literature on queuing applications [2,12,27,30,37,39], just to mention some of it. In the area of computer performance, see [27,37].

REFERENCES

[1]   Ajmone Marsan, M., Balbo, G., Conte, G. - *Performance Models of Multiprocessor Systems*, Computer Systems Series, MIT Press, 1986.

[2]   Allen, A. O. - *Probability, Statistics, and Queueing Theory with Computer Applications*, Academic Press, 1978.

[3]   Baskett, F, Chandy, K.M., Muntz, R.R., Palacios, F. - "Open, Closed and Mixed Networks of Queues with Different Classes of Customers," *Journal of the ACM* 22 (2), pp. 248-260, April 1975.

[4]   BEST/1 - Computer System Capacity Planning, BGS Systems, Inc., Lincoln, Massachusetts.

[5]   Beutler, F.J. and Melamed, B. - "Decomposition and Customers Streams of Feedback Networks of Queues in Equilibrium," *Operations Research* 26 (6), November-December 1978.

[6]   Burke, P. J. - "The Output of a Queuing System," *Operations Research*, pp. 699-704, 1966.

[7]   Burke, P. J. - "The Output Process of a Stationary M/M/s Queueing System," *Annals of Mathematical Statistics* 39 (4), pp. 1104-1152, 1968.

[8]   Buzen, J. P. - Queueing Network Models of Multiprogramming. Ph. D. Thesis, Division of Engineering and Physics, Harvard University, 1971.

[9]   Buzen, J. P. - "Computational Algorithms for Closed Queueing Networks with Exponential Servers," *Communications of the ACM* 16 (9), September 1973.

[10]   Chandy, K.M. and Neuse, D. - "Linearizer: A Heuristic Algorithm for Queueing Network Models of Computing Systems," *Communications of the ACM* 25 (2), February 1982.

[11]   Ching-Tarng, H. and Lam, S.S. - "PAM-A Noniterative Approximate Solution Method for Closed Multichain Queueing Networks," Technical Report TR-87-28, Department of Computer Sciences, The University of Texas at Austin, Austin, Texas.

[12]   Cooper, R. B. - *Introduction to Queueing Theory*, Second Edition, Elsevier North Holland, 1981.

[13]   Eager, D.L. and Sevcik, K.C. - "Bound Hierarchies for Multi-Class Queueing Networks," *Journal of the ACM* 33 (1), pp. 179-206, January 1986.

[14]   Garzia, R. F. - "Modeling the MVS/370 Computer Systems," 2nd International Symposium on Large Engineering Systems, University of Waterloo, Waterloo, Ontario, Canada, May 15-16, 1978.

[15]   Garzia, R. F. - "Queuing Model for the MVS/370 Computer Systems," Summer Computer Simulation Conference, Toronto, Canada, July 16-17, 1984.

[16]   Garzia, R. F. - "Computer Modeling - A Complete Case Study for the MVS Computer System," Technical Paper BR-1169, The Babcock & Wilcox Company, 1982.

[17]   Garzia, R. F. - "BWQUE - Interactive Program for Computer Performance Evaluation," 1983 Winter Simulation Conference, Washington, D.C., December 1983.

[18]   Gordon, W. J. and Newell, G. P. - "Closed Queueing Systems with Exponential Servers," *Operations Research* 15, pp. 254-265, 1967.

[19]   Hillier, F. S. and Lieberman, G. J. - *Introduction to Operations Research*, Holden-Day, 1986.

[20]   Jackson, J.R. - "Networks of Waiting Lines," *Operations Research* 5, 1957.

[21]   Jackson, J.R. - "Jobshop-like Queuing Systems," *Management Science* 10 (1), October 1963.

[22]   Jewell, W. S. - "A Simple Proof of $L = \lambda \ W$," *Operations Research 15*, pp. 1109-1116, 1967.

[23]   Kleinrock, L. - *Queueing Systems*, Volumes I and II, Wiley, 1975.

[24]   Konheim, A. G. and Reiser, M. - "Finite Capacity Queueing Systems with Applications of Computer Modeling," *SIAM Journal on Computing* 7, pp. 210-279, 1978.

[25] Kufin, P. - "Analysis of Complex Queuing Networks by Decomposition,"
      Proceedings of the 8th International Teletraffic Congress, Vol.1, Melbourne,
      Australia, November 10-17, 1976.

[26] Lam, S.S. - "Dynamic Scaling and Growth Behavior of Queueing Network
      Normalization Constants," *Journal of the ACM* 29 (2), pp. 492-513, April
      1982.

[27] Lavenberg, S. S. - *Computer Performance Modeling Handbook*, Academic
      Press, 1983.

[28] Lee, Y. - *Statistical Theory of Communications*, Wiley, 1960.

[29] Little, J. D. C. - "A Proof of the Queueing Formula $L = \lambda\ W$," *Operations
      Research* 9, pp. 383-387, 1961.

[30] Lazowska, E.D., Zahorjam, J., Graham, G.S., Sevcik, K.C. - *Quantitative
      System Performance: Computer System Analysis Using Queueing Network
      Models*, Prentice-Hall, 1984.

[31] Melamed, B. - "On Poisson Traffic Processes in Discrete State Markovian
      Systems with Applications to Queuing Theory," Technical Report 77-7,
      Department of Industrial and Operations Engineering, University of Michi-
      gan, 1977.

[32] McKenna, J., Mitra, D., Ramakrishnan, K. G. - "A Class of Closed Marko-
      vian Queueing Networks: Integral Representations, Asymptotic Expansions,
      and Generalizations," *The Bell System Technical Journal* 6 (5), May-June
      1981.

[33] McKenna, J. and Mitra, D. - "Asymptotic Expansions and Integral Represen-
      tations of Moments of Queue Lengths in Closed Markovian Networks,"
      Journal of the Association for Computing Machinery 31 (2), April 1984.

[34] Ramakrishnan, K. G. and Mitra, D. - "An Overview of PANACEA, a
      software Package for Analyzing Markovian Queueing Networks," *The Bell
      System Technical Journal* 61 (10), December 1982.

[35] Reiser, M. and Lavenberg, S.S. - "Mean Value Analysis of Closed Mul-
      tichain Queueing Networks," *Journal of the ACM* 27 (2), pp. 313-322, April
      1980.

[36] Sakata, M., Noguchi, S., and Oizuma, J. - "An Analysis of the M/G/1
      Queue with Round-Robin Scheduling," *Operations Research* 19, pp. 371-
      385, 1971.

[37] Sauer, C. H. and Chandy, K. M. - *Computer Systems Performance Modeling*,
      Prentice-Hall, 1981.

[38] Schmidt, P. H. and Garzia, R. F. - "Direct Calculation of the G Coefficients
      for the Generating Function of the Processing Network," *Computer Systems:
      Performance and Simulation*, edited by M. Ruschitzka, North-Holland, 1986.

[39] Stuck, B.W. and Arthurs, E. - *A Computer and Communications Network
      Performance Analysis Primer*, Prentice-Hall, 1985.

[40] Williams, A. C. and Bhandiwad, R. A. - "A Generating Function Approach
      to Queueing Network Analysis of Multiprogramming Computers," Chapter
      on Networks, Vol. 6, Wiley, 1976.

III
Wide Area
Communication Networks

7

Design and Performance Analysis of Survivable Networks

Clayton M. Lockhart

AT&T Bell Laboratories
Holmdel, New Jersey

1. INTRODUCTION

Robust networking is generally associated with packet-switched networks, following the work of Baran [3]. Because of the inherent routing flexibility of a nonhierarchical datagram network, theoretically it is possible for datagrams to find a surviving end-to-end path if one exists. Adaptive algorithms are often applied to datagram packet-switched networks [4,12], although adaptivity has been considered in circuit-switched networks [14,15]. The major problem with adaptivity as applied to common carrier circuit-switched networks is that the implementation of adaptive algorithms and information sharing can be costly. While various adaptive routing algorithms have been implemented in the somewhat experimental ARPANET [12], the long-distance carriers cannot afford to make major nonstandard modifications of switch software, development efforts that are incommensurate with the revenue derived from special customers representing typically less than 1% of the total traffic volume. Even in a private line network such as the AUTOVON [7], the switching and signaling technologies can limit the extent of flexible routing. Katz [11], in analyzing progressive alternate routing (with application to AUTOVON) and the means by which survivability could be extracted from a fixed routing plan, recognizes the importance of *physical diversity* as an important criterion in the design of routing tables.

A particular measure of physical diversity is developed in this chapter; however, physical diversity between routes can be loosely defined as their relative amount of environmental (spatial or frequency) separation. Diversity is not an absolute concept because its definition depends on the stress or threat to the network. For example, if stress is induced by optical fiber cuts, lack of diversity between routes is gauged by their physical coincidence, the route mileage along common rights of way. On the other hand, network stress may be caused by radio channel fading, in which case diversity may be obtained by employing differing frequencies or antennas on the same microwave towers. The nuclear targeting scenario threat discussed in this chapter leads to a measure of diversity which differs slightly from the fiber cut diversity measure.

In the context of modern circuit-switched networks, there is a need to revisit the concept of alternate routing based on physical diversity and to compare it against other network concepts such as adaptive routing and fixed, economic routing. This chapter motivates the need for and presents a method of designing robust, circuit-switched, virtual networks embedded in the Public Switched Network (PSN).

Major common carrier networks have evolved rapidly in the years just prior to and following the AT&T divestiture. Analog crossbar and step-by-step switching machines continue to be replaced by stored program controlled, digital switching equipment. At the highest levels of the major common carrier networks the switching machines will soon be digital, with most of them being able to serve large volumes of traffic. Because of traffic concentration to these large switching offices in or near metropolitan areas, the trunk groups serving the inter-city, long-haul demand are configured to interconnect each switch to nearly every other switch, resulting in a highly connected network. Carrier routing plans can make use of this high connectivity by exploiting traffic noncoincidence and variation in geographically distant parts of the network, a measure of flexibility not possible with hierarchical routing. Economic efficiency can be achieved by employing dynamic routing schemes such as the Dynamic Nonhierarchical Routing (DNHR) used in AT&T's network [2]. Networks can be protected from performing inefficiently under general and focused overload (or other forms of stress), by employing a variety of restrictive network management techniques. For instance, if certain central office codes are found to be "hard to reach" (HTR), then the codes are placed on HTR lists at various switches to block calls before they waste network resources in ineffective attempts to complete. To support the routing, network management, security and service features, an out-of-band signaling network is required, generally referred to as Common Channel (Interoffice) Signaling (CCS) [8].

Given the evolution of major common carrier networks, the possibility arises that an inexpensive robust network service can be implemented as a circuit-switched, software-defined network within individual inter-exchange carrier networks. The robust network design problem for special customers as applied to the PSN is solved for the most part by the implementation of appropriate routing tables and is to be contrasted with the standard method of procuring survivable networks by building dedicated, privately owned facilities or leasing private lines from common carriers. As long as the routing plan is based on physical diversity, the resulting network will have enhanced survivability with minimal software development and special hardware costs. With the proper network management features implemented, a customer would effectively have a survivable virtual private line network.

The chapter is organized as follows. Concepts required for understanding the network design algorithm are covered in Section 2, while the Robust Nonhierarchical Routing (RNHR) algorithm and an Economic Nonhierarchical Routing (ENHR) algorithm are discussed in Section 3. Survivability and diversity assessment of the robust and economic designs are presented in Section 4, and in Section 5 the carrier concerns of network implementation and network design maintenance associated with robust routing are discussed.

2. NETWORK CONCEPTS

To aid in the understanding of the RNHR algorithm we review the assumed network routing capability, the trunk and facilities networks, and their interrelation. The RNHR algorithm is based on certain notions of physical diversity which relate to the nature of network damage and congestion the network may experience in a stress situation. An overview of stress scenarios is necessary to see how RNHR helps overcome correlated damage and traffic congestion.

2.1 Network Topology, Routing and, Signaling

The switching and trunk group configurations of the circuit-switched networks considered here are generally termed "nonhierarchical" or "flat."

Each switch has the same routing capability and similar connectivity as the other switches. Because of the large call carrying and call set-up capacities of modern digital switching systems, major carriers are deploying large switches in their networks such that the network is nearly fully connected. The standard PSN hierarchy is being replaced by a two-tier network in which the lower tier consists of the Local Exchange Carriers (LECs) which connect to the upper tier of Interexchange Carriers (IXCs). The IXC networks are rapidly evolving into flat topologies because the switching points are placed in or near major metropolitan areas, among which there is enough traffic to often warrant direct trunk groups. Even the United Kingdom's digital switched long-distance network will be fully connected between about 50 major cities [13].

The network described here is representative of an IXC and will be assumed to route calls on paths consisting of either one or two trunks (links). Paths longer than two links are not considered, although some nonhierarchical robust networks, such as AUTOVON, use multi-link paths. Long-distance calls often use more than two links because the calls use the access and egress links in local exchange carrier networks. Any call uses either two or three switches, depending on whether it is routed over one link or two. A call enters the network at an originating switch (OS) and is destined for a terminating switch (TS). Adaptive routing is not considered here; the RNHR design algorithm pertains to a network which functions by an alternate routing scheme with one-link crankback, very similar functionally to AT&T's DNHR. The call might connect directly on the trunk group OS-TS or it might complete over a two-link route by using the via switches $\{VS_i\}$ through an alternate routing and a crankback capability as shown in Fig. 7.1. If call control is passed to a via switch VS and the VS-TS trunk group is either totally busy or damaged, the call control will be referred back to OS, a procedure known as crankback. The call will then be alternate routed to the next via switch. This kind of route control allows a thorough exploration of the network without having a large number of links ultimately involved in the final call configuration. A progressive routing mechanism (as in the old hierarchical routing) would block the call attempt when the VS-TS trunk group was found to be busy. The more highly connected the network is to begin with, the more successful a two-link routing mechanism is in searching out surviving routes.

Depending on the IXC, various network management controls exist to protect trunk group, switching system, and signaling network resources. The implementation of such a nonhierarchical routing scheme with network

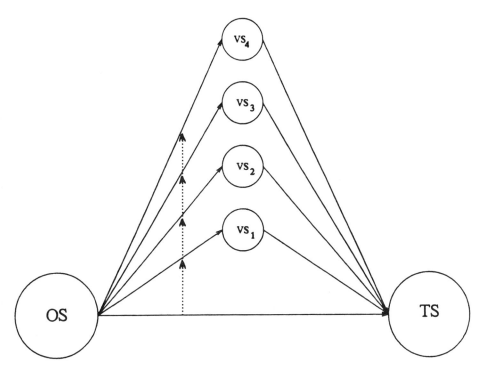

Figure 7.1 Alternate Routing Plan (with Crankback) Used in the Nonhierarchical Circuit-Switched Network.

management controls requires computer-controlled switching systems not only to perform the aforementioned routing functions but to accumulate network statistics, determine if threshold criteria have been exceeded and to apply network management controls. Out-of-band CCS is generally employed to pass call status information, crankback and failure messages, traveling class marks, and network management signals, all of which are useful for modern routing and network management capabilities.

2.2 Telecommunications Databases

While databases might be thought to be a peripheral topic relative to design algorithms, it turns out that the heart of the RNHR algorithm makes inti-

mate use of the trunk and facilities databases. Without such databases the RNHR algorithm would be useless. The required databases are listed as follows:
- Switches
- Trunks
- Facilities
- Spans
- Circuit layout (trunk-to-facility relation)
- Facility-to-span relation

The switching system database should consist of those switches capable of the kind of routing and signaling previously described. Coordinates and some sort of naming convention should be included. Typically, switches are labeled by Common Language Location Identification (CLLI) codes, with coordinates taken from the Continental United States Cartesian Vertical and Horizontal (V-H) coordinate system. Trunks are analog or digital communications channels between switches. A trunk group is identified by the CLLI codes of the switch endpoints, the number of trunks in the trunk group, the method of signaling, and the trunk group's usage in the routing plan. Each trunk is further identified by a trunk number.

The term "facilities" refers to the multiplex facilities in the analog and digital multiplexing hierarchy [5]. Even though the IXCs are rapidly expanding their digital facility networks, analog facilities will persist for some time. For the model network considered here, we will consider only those facilities at the lowest level of multiplex which are the analog channel group (12 voiceband channels) and the digroup (1.544 Mbps, commonly referred to as a "T1" facility, equivalent to 24 voiceband channels). Facilities are identified by their terminus CLLI codes, multiplex level, and a facility number (similar to a trunk number). Groups and digroups are the facilities involved in the circuit layout database, which is the relation between trunks and the first level of multiplex. Circuit layout relates each trunk to the sequence of groups or digroups into which the trunk is multiplexed.

Spans are basically facilities at the highest level of multiplex, with cross sections typically in the tens of thousands of trunks. Spans have the finest physically resolution: while a span may average about 30 miles in length, a channel group averages about 500 miles in length. The terms "optical fiber" or "digital radio" refer to facilities at the span level of multiplex. Adequate physical resolution is necessary to properly assess damage and to calculate physical diversity. With linked databases describing the multiplex hierarchy it is possible to create a facility-to-span database which relates channel groups and digroups to spans. From the circuit layout database it is then possible to relate trunks to spans through the relation with facilities. An example of the relation among the databases is shown in Fig. 7.2. A single trunk group (shown to have 24 trunks) is multiplexed on two analog channel group facilities routed over different physical span paths.

2.3 Customer Requirements

The customer traffic matrix is often difficult to ascertain, especially for a robust network service whose main utility manifests itself during a time of network stress. In designing a new network service, usually the service concept is presented to prospective customers as a market survey. Traffic models are then estimated on the basis of the nature of the service and previous experience in calling patterns. In the absence of any calling patterns just the number of lines served behind each switch and the average line

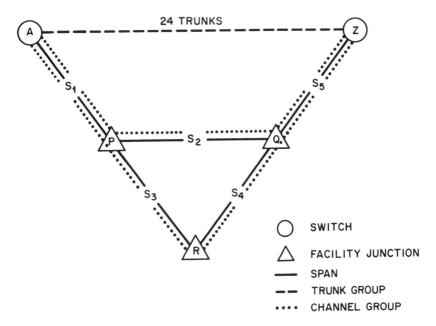

Figure 7.2 Sample Circuit Layout for Trunk Group A-Z Showing the Relation to Facilities.

usage will help define a traffic model. The traffic matrix, while interesting for determining revenues from the robust network service, does not greatly influence the robust network design philosophy as presented here. Because the critical traffic volume is assumed to be very small with respect to the public traffic (perhaps 1%), any change in the demand can be easily accommodated. This holds true except when traffic sensitive hardware is required for the critical customer, for instance, transmission treatment circuits for data calls. In this case, an increase in the demand requires a proportionate increase in such hardware. As long as the network remains software-defined, the design will be relatively insensitive to changes in demand.

There are three basic grades of service (GOSs) to be considered in the type of network discussed in this chapter: call completion probability, call set-up time, and the average number of trunks (related to transmission quality). Call completion probability is generally the most important GOS, and it will be the measure by which the performance of the network is monitored under damage, the results of which are discussed in Section 4. Even though more alternate routing can take place in a damaged network, because of the fast signaling associated with call set-up, the resulting delay will not be significant to the customer. Transmission quality will likewise be relatively high because routing is restricted to two trunk groups (plus standard access and egress through local carriers). If the routing had required more trunk groups in tandem, it is conceivable that transmission compensation might be required to counteract loss and envelope delay distortion.

2.4   Stress Scenarios

Network stress can be categorized into damage effects and traffic congestion. Damage generally is a long-time scale phenomenon, with repair taking place hours and even up to weeks after the damage has occurred. Coaxial cable or optical fiber repair may take hours, while a large switching office fire repair operation may take weeks. In a nuclear attack situation repair may only occur after months or not at all. Pure traffic congestion effects are much more transient phenomena, especially with the employment of modern network management techniques. Focused overloads can be handled quickly by blocking calls to HTR codes, while general overloads can be handled by call gapping, dynamic overload control and trunk reservation techniques [8]. Damage and congestion effects are often related. If the network suffers some major outage, congestion might appear due to a multitude of ineffective attempts to stations near the afflicted area. A global congestion problem may occur in a limited attack because the population would be left relatively unharmed while the network may suffer outages. On the other hand, if a massive nuclear attack scenario occurred, such as the Federal Emergency Management Agency (FEMA) scenario [10], not only would the network suffer damage but the number of network users would decrease (due to the counter-population targeting), a situation in which it is not clear a priori if network congestion would exist. Lesser damage scenarios [1], in which weapons are targeted against nuclear forces and the communications supporting those forces, would leave more of the population and network intact, with the result that the network could be subjected to overloads.

We will briefly describe the damage mechanism and the way in which trunk group survivability is assessed as a function of physical component damage. For attack scenarios we assume that any span endpoints (microwave towers, central offices, etc.) that lie within a well-defined afflicted area are destroyed, a damage assessment process often referred to as a "cookie cutter." The radius of the damage zone (the cookie cutter "kill radius") depends on the yield and height of burst (HOB) of the nuclear weapon and also on the blast hardness and brittleness of the communications facility. Referring to the trunk group $A - Z$ shown in Fig. 7.2, suppose a weapon destroys point R. The trunk group survives but with a diminished capacity of 12 trunks. If either of the points Q or P is destroyed, the whole trunk group is destroyed. The survivability of the trunk groups is determined through the combination of damage assessment at the physical level and the relational databases. Such a damage assessment process produces *correlated damage* for the trunks in the trunk group, the correlation arising from the multiplexing relation of trunk groups to channel groups and channel groups to spans, as reflected in the relational databases. On the average, the longer the trunk group, in terms of the number of nodes and spans it uses, the more vulnerable it is to collateral damage. The problem of network design in a situation of correlated damage has been discussed before [6,9]. This chapter provides a realistic description of the correlation properties and a network design algorithm which counters the effects of correlated damage.

To assess the performance of RNHR under damage, we consider a sequentially optimal targeting scheme employing cookie cutter damage assessment as applied to a model network with 97 switching centers. Each of the switching centers is directly targeted by a one-megaton weapon at the optimal HOB to achieve the maximum kill radius for targets that have a threshold hardness of about 2 lb per square inch (psi) to an overpressure shock wave. According to [16], the optimal HOB is about 10,000 ft with a resulting 2 psi kill radius of 42,000 ft or about 8 miles. For this model, any telephone plant (span endpoints) within 8 miles of a switch coordinate is assumed to be destroyed while anything outside this radius survives, consistent with the cookie cutter methodology. The 97 targets are sequentially ordered according to each target's successive effectiveness in reducing the total number of trunk groups in the network. Given that n targets had been hit, the target $n + 1$ is picked from the $97 - n$ remaining targets by finding that target which reduces the number of surviving trunk groups by the largest amount. Because a building in which a switch is housed is generally used as an intermediate multiplexing point for trunk groups which do not terminate at the switch, it is possible to destroy many trunk groups without having to destroy their terminating switches. One can see the importance of the relational databases for determining proper damage assessment. For a given number of targets hit, the ratio of destroyed trunk groups to the original undamaged number will be referred to as the *fraction of network damaged.*

2.5   A Diversity Measure

The physical diversity of routes between two switching offices can be a difficult measure to define, especially if the routes involve multiple trunk groups. An often asked question is how many physically distinct routes exist between two switching offices. An analysis of logical diversity (switching office diversity) is not enough to answer the question. Generally, some sort of node or arc (span) cut-set analysis is necessary, but even then the analysis is not complete without taking into account restrictions imposed by the relation between trunk groups and spans. It is not possible to route with arbitrary flexibility on the span facilities because of multiplexing restrictions. Because the networks considered here route calls on one or two links between an origination-destination pair of switching offices, it is simple to define the relative diversity of a route with respect to another route or collection of routes.

An example route configuration is shown in Fig. 7.3. The two switching offices for the two routes are A and Z, while the two-link routes, $A - V_1 - Z$ and $A - V_2 - Z$, are represented by the dashed lines directly between the switches. Spans are represented by the segments $s_1, \ldots, s_{12}$, and the triangles represent facility junctions or major power feed points,

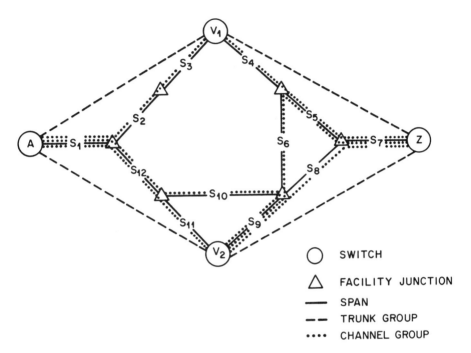

Figure 7.3 Sample Via Routes Between Switches A and Z Showing the Relative Diversity of Routes from a Facility Perspective.

not the repeaters in a large coaxial cable which can occur every mile. The relation between the logical and physical is determined by the multiplexing equipment configuration. The physical routing of the trunk groups on the spans is determined by the channel groups (dotted lines) in Fig. 7.3. The circuit layout and channel group-to-span relation for the four trunk groups is given in Table 7.1.

Note that the two-link route $A - V_2 - Z$ employs channel groups over two different span sequences, a typical occurrence of trunk group splitting in the network. The two routes $R_1 = A - V_1 - Z$ and $R_2 = A - V_2 - Z$ are logically diverse from each other because $V_1 \neq V_2$. However, they are not physically diverse, as is clear from the figure. A *path* $P(R)$ corresponding to a route R between switches A and Z (either one trunk group or two trunk groups in tandem) is defined by the set of spans used by the trunk

Table 7.1   Trunk Group to Facility Relation

Trunk group	#Trunks	Channel group	Facility #	Span sequence
$A-V_1$	12	$A-V_1$	#1	$s_1 - s_2 - s_3$
V_1-Z	12	V_1-Z	#1	$s_4 - s_5 - s_7$
$A-V_2$	24	$A-V_2$	#1	$s_1 - s_{12} - s_{11}$
		$A-V_2$	#2	$s_1 - s_{12} - s_{10} - s_9$
V_2-Z	24	V_2-Z	#1	$s_9 - s_8 - s_7$
		V_2-Z	#2	$s_9 - s_6 - s_5 - s_7$

group(s) comprising R. From the example shown in Fig. 7.3, $P(R_1)$ is given by the set $s_1, s_2, s_3, s_4, s_5, s_7$ while $P(R_2)$ is given by the set $s_1, s_5, s_6, s_7, s_8, s_9, s_{10}, s_{11}, s_{12}$. For a set of spans S, let $n(S)$ denote the number of unique spans in S. The amount of *overlap* $O(P_2|P_1)$ of a path $P_2 = P(R_2)$ with respect to $P_1 = P(R_1)$ is defined by

$$O(P_2|P_1) = n(P_1 \bigcap P_2)/n(P_2) \qquad (7.1)$$

Note that the measure is not symmetric, i.e., $O(P_j|P_k) \neq O(P_k|P_j)$, in general. The overlap measure applies to unions of paths, a property which is used in the RNHR design algorithm. The overlap measure of a path P and a union S_{AZ} of a set of paths, $P_1 \bigcup \cdots \bigcup P_k$, where all paths are between switching points A and Z is given by

$$O(P|S_{AZ}) = n(P \bigcap S_{AZ})/n(P) \qquad (7.2)$$

For the example in Fig. 7.3, the calculations of O yield $O(P_2|P_1) = 0.333$, because P_2 has 3 of its 9 spans in common with P_1. An inspection of Fig. 7.3 gives one the impression that the amount of span overlap between the two (dashed) routes, $A-V_1-Z, A-V_2-Z$, as physically laid out over the (dotted) multiplex facilities, is about one-third.

The path overlap formula, Equation (7.1), can be redefined so as to be compatible with the nature of the network stress, as discussed briefly in the introduction. The overlap measure reflected in Equation (7.1) is derived from the damage assessment process in which path vulnerability is related to the number of span endpoints. Path survivability is determined by the cookie cutter method applied against span endpoints. Another network threat of a commercial interest to interexchange carriers is that of optical fiber cuts. Span vulnerability under this threat (average number of cuts over a time period) is proportional to its length. The overlap function might be defined by

$$O(P_2 | P_1) = d(P_1 \cap P_2)/d(P_2) \qquad (7.3)$$

where $d(P)$ is the sum of the span mileages for the unique spans in a path P. This measure of overlap introduces a distance bias so that paths with more span mileage in common are less diverse, independent of the number of span endpoints involved.

Consider a set of via routes $\{A - V_i - Z\}$ between switches A and Z, the associated paths $\{P_i = P(A - V_i - Z)\}$, and a set S_{AZ} of paths between switches A and Z. The diversity $D(P_i | S_{AZ})$ of P_i with respect to S_{AZ} is defined by

$$D(P_i | S_{AZ}) = 1 - O(P_i | S_{AZ}) \qquad (7.4)$$

3. DESIGN ALGORITHMS

In this section we give a high-level description of two routing table design algorithms, RNHR and Economic Nonhierarchical Routing (ENHR). RNHR computes routes based on the notions of physical diversity covered in the previous sections, while ENHR ignores physical diversity (while retaining logical diversity) and merely computes routes by determining the shortest two-link paths. One should keep in mind that these algorithms are not used for adaptive, real-time routing computations in switching machines. They are used to produce predetermined routing plans implemented in switches by the common carriers and represent network designs for special customers.

3.1 RNHR Algorithm Description

Consider two switches A, Z chosen from the set of N switches in the network. For all pairs (A, Z), the RNHR algorithm determines the routing table used for calls entering at A and exiting at Z. The routing table is of the form:

$\underline{A \rightarrow Z}$

$\quad R_1$
$\quad R_2$
$\quad \cdot$
$\quad \cdot$
$\quad \cdot$
$\quad R_n$

For the pair (A,Z) define $n(A,Z)$ to be the number of via switches $\{V_i\}$ such that the two-link routes $\{A - V_i - Z\}$ exist in the trunk group network. If the trunk group $A - Z$ exists in the network, then $R_1 = Z$. The subsequent switches $\{R_2,\ldots,R_n\}$ are the via switches for calls routed between A and Z.

RNHR ALGORITHM

SET PARAMETERS:

Specify a triple (n,Δ,δ), where n is the number of routes on the RNHR tables, Δ is the desired diversity criterion, and δ is a diversity decrement factor constrained such that $m\delta = \Delta$, where m is a positive integer. Calculate the airline distances $\{d(A,Z)\}$ between switches A and Z for all pairs (A,Z). Here, the integer n is chosen to be a constant over all pairs (A,Z). n could be allowed to vary with (A,Z) by choosing it in some fashion within Step 1.

START Pick a pair (A,Z) for which a routing table is yet to be determined. If no such pair remains, go to END.

STEP 1 Determine the candidate via switches $\{V_i\}$ such that the two-link routes $\{A - V_i - Z\}$ exist in the trunk group network. Calculate the values $\{M_i = d(V_i, M_{AZ})\}$ where M_{AZ} is the midpoint between A and Z. Rank order the candidate switches $\{V_i\}$ according to their values $\{M_i\}$, and relabel the switches such that those with the lowest labels have the lowest values.

STEP 2 Initialize an array $VALUE[\ ] = 1$ (array length N), and $\bar{\Delta} = \Delta$. If the route $A - Z$ exists, initialize the set S_{AZ} to contain the path $P(A - Z)$, and set $R_1 = Z$. Otherwise, initialize S_{AZ} to contain the path $P(A - V_1 - Z)$, and set $R_1 = V_1$. Set $j = 2$.

STEP 3 Set $i = 1$.

STEP 4 If $VALUE[V_i] < \bar{\Delta}$, go to Step 7.

STEP 5 Compute the diversity measure $D = D(P_i | S_{AZ})$, where $P_i = P(A - V_i - Z)$. If $D < \bar{\Delta}$, then P_i is not diverse enough relative to S_{AZ}, set $VALUE[V_i] = D$, and go to Step 7. If $D \geqslant \bar{\Delta}$, then P_i is sufficiently diverse from S_{AZ}, include P_i in the set S_{AZ}, set the node $R_j = V_i$, set $VALUE[V_i] = 0$, and increment j by one.

STEP 6 If $j = n + 1$, the algorithm is finished for the pair (A,Z). In this case, print the set $\{R_i\}$ to create the (A,Z) routing table and return to START.

STEP 7 Increment i by one. If $i \leqslant n(A,Z)$, go to Step 4. Otherwise, the number of via candidate switches have been exhausted before the routing table has been filled, in which case, decrement $\bar{\Delta}$ by δ, and go to Step 3.

END

The network designer may want the RNHR algorithm to produce routing tables reflecting physical diversity in an undamaged network, but in addition for it to produce routes which preferably survive a given damage scenario, a requirement called "tailoring." To achieve tailored RNHR, Step 1 should be modified so that the set of candidate via switches $\{V_i\}$ is determined by the existence of two-link routes $\{A - V_i - Z\}$ in the *surviving* trunk group network. Before running the tailored RNHR algorithm, the damage scenario must be applied to the relational databases to determine which trunks groups survive. The algorithm is then the same except for Step 7, in which $\bar{\Delta}$ is *not decremented* by δ, with control not returning to Step 3. At the conclusion of the algorithm, incomplete routing tables containing surviving routes having mutual diversity of at least Δ will have been created. The RNHR algorithm can then be run on the undamaged network ("untailored" RNHR) to fill out the tables to length n. Generally, Δ should be set lower for the tailored RNHR algorithm than for the untailored RNHR. Because of the damage incurred in the network, it is difficult to achieve the same degree of route diversity as is possible in an undamaged network.

The way in which RNHR is tailored to the optimal targeting scenario is slightly different as follows. For each pair (A,Z) calculate the untailored RNHR table until considering the choice of the final route R_n. Determine the fraction of network damage f_{AZ} such that at least one two-link route exists, but such that no routes survive beyond this level of damage. f_{AZ} is the "damage threshold" for the pair (A,Z). Randomly select one of the surviving two-link routes at the damage threshold, say V. If this route is not contained in the set $\{R_1,...,R_{n-1}\}$, then set $R_n = V$, or else set R_n equal to what the untailored RNHR algorithm yields for the final route. In this way at least one route on the (A,Z) RNHR table will survive at the damage threshold f_{AZ}.

One might question the rank-ordering procedure of Step 1 as to whether it ultimately leads to the best choice of the switches $\{R_i\}$. After all, there are $n(A,Z)!$ choices of ordered lists of via switches. The RNHR algorithm has an order dependence in that the choice of R_{j+1} depends on the switches $R_1,...,R_j$ chosen previously; therefore, the initial rank-ordering procedure has an effect on the routing table outcome. One observation is that, even though there are many *logical* routes to choose from, there are a limited number of *physically diverse* routes available. The physical network is the overwhelming constraining factor in determining the appropriate logical routes to put in a routing table. It might be hypothesized that any reasonable initial rank-ordering scheme for the via switches will lead to RNHR tables of comparable overall diversity and to about the same performance under damage. RNHR's use of shorter routes first is desirable because of their lessened vulnerability to collateral damage, as discussed in Section 2. The reason shorter routes are calibrated from the $A-Z$ midpoint is that a "bowing" effect is created in which the via switches are spread out somewhat evenly from the midway point. If the distances $\{d_{AZ}^i\}$ are used instead to rank order via switches (as in ENHR), the via switches can often cluster near one another, especially if one of the points A or Z is in a region of high population density. This clustering of via points is undesirable (for instance, if some regional disaster occurred and overload to several switches in the area existed), even if the routes involved meet the physical diversity criterion.

The RNHR algorithm does not guarantee the optimal setting of (n,Δ,δ), yielding a maximum level of diversity or the best performance under damage. In the performance analysis section the levels of diversity and survivability obtained by different parameter triples (n,Δ,δ) are shown.

3.2 ENHR Algorithm Description

The ENHR algorithm is a simple procedure to determine routing tables for each switch pair such that the calls will be routed on least cost routes. Given the switch pair (A,Z) and the candidate via switches $\{V_i\}$ as defined previously, compute the distances $\{d_{AZ}^i = d(A,V_i) + d(V_i,Z)\}$. Rank order the candidate switches $\{V_i\}$ and relabel them such that switches with lower indices have correspondingly lower distances $\{d_{AZ}^i\}$. To create the routing table, simply put the node Z (if the trunk group $A-Z$ exists) on the (A,Z) table and the highest ranked candidate via switches to fill out the table to length n. The routing choices are then logically diverse and will yield lower costs in terms of facility usage; however, facility diversity is not guaranteed. In the section on performance analysis it will be shown how much physically diverse routing is obtained by ENHR.

There is no close relation between ENHR and AT&T's DNHR [2]. While the DNHR algorithm attempts to minimize facility usage in a way similar to ENHR, it encompasses many other factors such as public message traffic loads, time-of-day variations, blocking criteria, and trunk group sizing, as well as an adaptive path selection capability implemented through a centralized network management system.

4. PERFORMANCE ANALYSIS

There are many different ways of expressing network survivability. In general, survivability is measured by the grade of service (GOS) the network offers to its users in a stress condition. To assess the performance of the RNHR and ENHR designs, combinations of damage and traffic congestion effects are used to stress the network, and various GOSs are monitored. In addition, a route diversity comparison between RNHR and ENHR is made. The ENHR and RNHR algorithms were run for a 97-node network, the nodes having an average trunk group degree of 88 (representing a graph that is about 92% connected). Table 7.2 shows the cases run and the diversity measures used in computing routing tables. The case in which tailoring is used refers to the tailoring associated with the optimal targeting scenario.

4.1 Physical Diversity Comparison

We show the results of the runs for the (A,Z) pair of Akron and Albuquerque. Fig. 7.4 shows the logical and physical connectivities for ENHR (Case 4) routing while Fig. 7.5 shows the RNHR routing (Case 1). Clearly, RNHR exploits a much more extensive range of facilities than ENHR for this switch pair, even though both routing tables have 14 routes.

A network-wide comparison of the relative physical diversity between ENHR and RNHR is possible by keeping track of the physical diversity each successive route possesses relative to the previous contents of the table. If the (A,Z) table has routes $\{R_i\}$, then the diversity D_i of R_i relative to the previous routes is given by $D_i = D(P(R_i) \,|\, S_{AZ}^i)$, where

$$S_{AZ}^i = \bigcup_{j=1}^{i-1} P(R_j), \qquad i = 2,\dots,n \tag{7.5}$$

	Table 7.2	Cases Run			
Case #	Routing	n	Δ	δ	Tailoring
1	RNHR	14	.90	.10	No
2	RNHR	14	.90	.10	Yes
3	RNHR	14	.75	.25	No
4	ENHR	14	-	-	-
5	ENHR	5	-	-	-

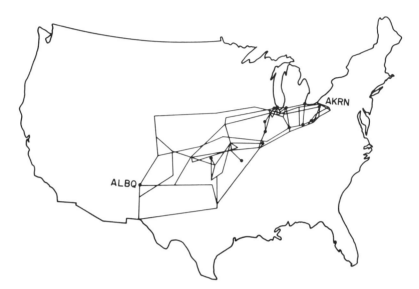

Figure 7.4   Facility Perspective for ENHR Routes (Case 4) Between Albuquerque and Akron.

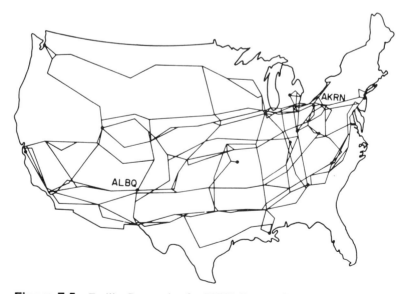

Figure 7.5   Facility Perspective for RNHR Routes (Case 1) Between Albuquerque and Akron.

S^1_{AZ} is defined to be the null set ϕ so that $D_1 = 1$. For a given pair (A,Z) and the number r $(r = 1,...,n)$ of routes on the routing table, the *effective number of routes* $R(r)$ is defined by

$$R(r) = \sum_{i=1}^{r} D_i \qquad (7.6)$$

Fig. 7.6 shows $R(r)$ as a function of r for the RNHR and ENHR cases run. Clearly, RNHR possesses more inherent facility diversity than ENHR, as corroborated by Fig. 7.4 and 7.5. The curves suggest that the answer to the question, "How many physically diverse routes are there between two switches?" has a soft answer in that there is no clear "knee of the curve" or saturation point. Physical diversity is not calculated by a logical connectivity analysis, in which case the answer is 14 (i.e., the size of the routing table), nor by a span min-cut set analysis, which would yield a span cutset of two spans by inspection of facility maps not presented

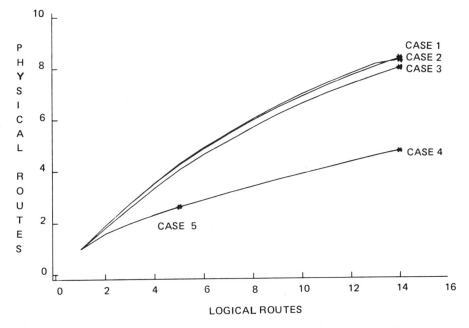

Figure 7.6 Effective Number of Physically Diverse Routes Versus the Number of Logical Routes (Length of Routing Tables).

here. The answer depends on the routing scheme (RNHR or ENHR), the trunk group connectivity, and the relational databases to the span level.

4.2   Performance Under Damage

Besides RNHR and ENHR we introduce two additional routing schemes for the purposes of comparison: ideal 2-link routing, and ideal routing. Ideal 2-link routing manages to find end-to-end connectivity in the trunk network, if it exists at all, subject to the 2-link routing restriction similarly imposed on ENHR and RNHR. Ideal 2-link routing represents what a perfect adaptive routing scheme could render and still not violate the standard routing method of the common carrier. Ideal routing, on the other hand, finds any existing end-to-end connectivity in the trunk network, with no link restriction imposed. Ideal routing, while useful for finding connectivity, in high damage situations can present a transmission quality problem because the number of trunks per call can be high for a significant fraction of calls.

Relative connectivity is defined to be the ratio of the point-pair connectivity for a given routing scheme relative to the point-pair connectivity for the ideal routing scheme. As a function of the fraction of the network damaged, the relative connectivity measures how close to ideal a particular routing scheme is performing. The relative connectivities of the RNHR cases, ENHR, and the ideal 2-link routing schemes are shown as functions of network damage in Fig. 7.7. Note that untailored RNHR (Case 1) comes fairly close to the ideal 2-link routing, and that it significantly outperforms five route ENHR (Case 5). The tailored RNHR performance is identical to the ideal 2-link routing because of the tailoring of the fourteenth route. The implication is that tailoring or adaptivity is useful only at the highest levels of damage (70%–100%). Alternate routing performs well in low and medium levels of damage as long as the routes are chosen on the basis of physical diversity. The performance of tailored RNHR degrades somewhat relative to a perfect network at the highest damage levels because at some point the trunk group network is damaged to such a degree that the ability to route some calls over more than two links is necessary to achieve end-to-end connectivity.

The relative connectivity results can be interpreted as measures of relative call completion probabilities given that calls experience no blocking on individual surviving trunk groups. If network management is very effective, and the critical calls are exempt from network management controls, the curves shown in Fig. 7.7 are adequate for determining the relative effectiveness of routing schemes in terms of call completion probabil-

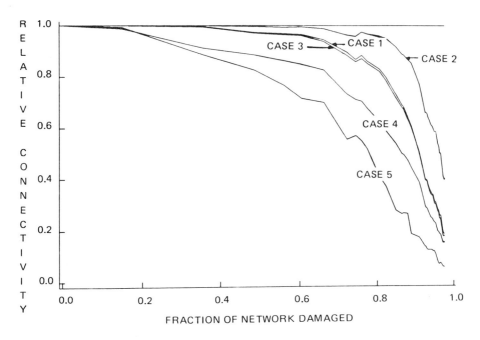

Figure 7.7 Connectivities of Various Routing Plans (Relative to Perfect Routing) Versus Damage Level. Case 2 (Tailored RNHR) and the Perfect 2-Link Routing Have Identical Performance for This Measure.

ity. On the other hand, if a call experiences trunk group blocking, then the number of surviving routes becomes an important factor in determining call completion. Relative to those switch pairs for which at least one surviving route exists, the average number of surviving routes is plotted against network damage in Fig. 7.8. The bunching of the curves implies that the effectiveness of RNHR is not so much that it provides more routes but that it enhances the possibility that *any routes are available,* as evidenced by Fig. 7.7. Tailoring also helps provide more surviving routes as well as higher connectivity in the worst case of optimal targeting. Tailoring, by the nature of its scenario dependence, can be fragile, that is, at high damage levels perturbations in the scenario will obliterate the tailored routes. The RNHR design philosophy is to choose a number of routes on the basis of physical diversity, so that routing tables are for the most part untailored. A network possessing the inherent route diversity of RNHR

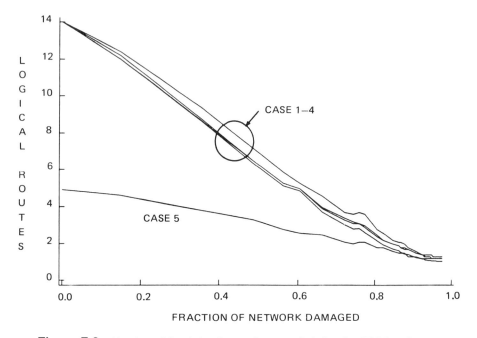

Figure 7.8   Number of Surviving Routes Between Switches for Which at Least One Route on the Routing Table Survives Versus Damage Level.

has the potential to cope with various damage effects and traffic congestion, although much of the traffic congestion problem is not solved by the robust routing inherent in RNHR but by the network management controls placed on the network.

5. CONCLUSION

It has been shown that for a highly connected commercial switching network a robust network design algorithm exists which produces designs with point-pair connectivity approaching that of a network having arbitrarily flexible routing. A network using RNHR significantly outperforms a network using economical routing even though both networks are constrained to use the same routing mechanism. It is shown that adaptive routing or scenario tailoring of RNHR tables is useful beyond a high level of network damage. The RNHR algorithm makes explicit use of relational network databases and produces a routing plan that counters the effects of correlated damage typical of complex communications networks.

There are a number of practical issues to be considered in the actual implementation of the RNHR algorithm. Use of RNHR requires some means of indicating to the switches involved in call set-up that the call is an RNHR call. This indicator can generally be presented as non-standard digits to the switches (such as a special area code attached to the normal called party address). The switches must be able to recognize the·digit sequence and translate it into an RNHR alternate route table. The translation process requires addition routing data memory as well as memory addressing capacity, switch resources which could be crucial, depending on the switch hardware and software parameters.

Even though the RNHR design is installed by downloading RNHR routing tables into switch memory, the design process is never complete and must continue as a maintenance procedure. Several factors conspire to make any design obsolete or degrade through time. Not only is the network in a state of constant flux, but the threat scenarios change with time. Facilities constantly change due to planning and provisioning processes carried out by common carriers. The assignments of channel groups and digroups to spans change because of economic analyses and newly deployed facilities. The assignments of circuits to channel groups (circuit layout) change because of the forecasting process associated with the sizing of trunk groups. One can think of this churn factor as a kind of "administrative damage" to the RNHR design. Because the RNHR algorithm produces routing patterns which are robust to damage, the effect of churn on the end-to-end performance of the RNHR service will be minimal. However, to keep the design from degrading significantly, an RNHR redesign would be necessary on a periodic basis.

As major carriers retire analog facilities and deploy new digital facilities, the physical networks will become sparse but will have large cross sections, implying increased outage risk. While any reduction of physical facilities will clearly reduce the effectiveness of RNHR to outage (as well as that of ENHR), the vulnerability of a thin, high capacity network points to the need in the message traffic segment for either robust routing or a rapid facility restoration capability. There is a trade-off between the costs associated with the facility route mileage required for RNHR (above that of ENHR-type routing) versus the risk of revenue loss in the use of non-robust routing.

The RNHR algorithm can be used for highly connected virtual circuit data networks in addition to the circuit-switched network discussed here.

Physically diverse virtual circuits can be set up by routing call set-up packets along logical paths which are mutually diverse. For the RNHR algorithm to apply directly, the trunk group connectivity of the data network
would have to be fairly high, so that all virtual circuits could be set up on
one or two-link paths.

REFERENCES

[1]   Arkin, W. M. and Fieldhouse, R. W. - *Nuclear Battlefields,* Ballinger Publish-
      ing Company, pp. 89-92, 1985.
[2]   Ash, G. R., Cardwell, R. H., Murray, R. P. - "Design and Optimization of
      Networks with Dynamic Routing," *Bell System Technical Journal* 60 (8), pp.
      1787-1820, 1981.
[3]   Baran, P. - "On Distributed Communications Networks," *IEEE Transactions
      on Communications Systems* COM-12, pp. 1-9, 1964.
[4]   Boorstyn, R. R. and Livne, A. - "A Technique for Adaptive Routing in Net-
      works," *IEEE Transactions on Communications* COM-29 (4), pp. 474-480,
      1981.
[5]   *Engineering and Operations in the Bell System.* Second Edition, R. F. Rey,
      Technical Editor, AT&T Bell Laboratories, Murray Hill, New Jersey, 1984.
[6]   Frank, H. - "Survivability Analysis of Command and Control Communica-
      tions Networks--Part I," *IEEE Transactions on Communications* COM-22 (5),
      pp. 589-595, 1974.
[7]   Gorgas, J. W. - "The Polygrid Network for AUTOVON," *Bell Laboratories
      Record* 46 (7), July/August 1968.
[8]   Haenschke, D. G., Kettler, D. A., Oberer, E. - "Network Management and
      Congestion in the U. S. Telecommunications Network," *IEEE Transactions on
      Communications* COM-29 (4), pp. 376-385, 1981.
[9]   Heffes, H. and Kumar, A. - "Incorporating Dependent Node Damage in
      Deterministic Connectivity Analysis and Synthesis of Networks," *Networks*
      16, pp. 51-65, 1986.
[10] *High Risk Areas.* Federal Emergency Management Agency Document TR-
      82, September 1979.
[11] Katz, S. S. - "Alternate Routing for Nonhierarchical Communication Net-
      works," Proceedings of the IEEE 1972 National Telecommunications
      Conference (NTC '72), pp. 9A-1-9A-8, Dec. 1972.
[12] McQuillan, J. M., Richer, I., Rosen, E. C. - "The New Routing Algorithm
      for the ARPANET," *IEEE Transactions on Communications* COM-28 (5),
      pp. 711-719, 1980.
[13] Mees, A. - "Simple is the Best for Dynamic Routing of Telecommunica-
      tions," *Nature* 323, p. 108, September 1986.

[14] Narendra, K. S., Wright, E. A., Mason, L. G. - "Application of Learning Automata to Telephone Traffic Routing and Control," *IEEE Transactions on Systems, Man, and Cybernetics* SMC-7 (11), pp. 785-792, 1977.

[15] Srikantakumar, P. R. and Narendra, K. S. - "A Learning Model for Routing in Telephone Networks," *SIAM Journal on Control and Optimization* 20 (1), pp. 34-57, 1982.

[16] *The Effects of Nuclear Weapons.* Third Edition, S. Glasstone and P. J. Dolan, Eds., The U. S. Department of Defense and the Energy Research and Development Administration, 1977.

8

A Study of Network Adaptive Routing

Mario R. Garzia

AT&T Bell Laboratories
Columbus, Ohio

1. INTRODUCTION

The behavior of distributed, nonhierarchical networks operating under localized congestion and node outage (damage) is at present not well understood. In this chapter we present a study of the use of local and global adaptive routing to improve network performance under stress.

The networks that we consider give their customers a large amount of freedom through the use of non-hierarchical alternate routing, crankback, and other network controls such as precedence. It has been previously noted [1] that providing such power to all users is not always desirable. The reason for this is that, with the use of all these controls, calls remain in the network a long time whether they eventually complete or fail. Depending on the state of the network, congestion and outages, some calls will have a very low probability of completion but still use all its capabilities in trying to set up the call, thereby using, for a long period of time, facilities that could be used more effectively to set up other calls. On the other hand, it can be shown [2] that the use of such powerful controls for a small portion of the network traffic, high-priority traffic, improves considerably the performance of this group of calls. Of course, this occurs at the expense of the low priority-traffic. For our discussion, however, we consider networks with alternate routing and crankback capabilities and inves-

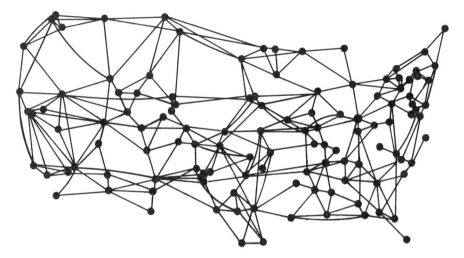

Figure 8.1   120-Node Sample Network.

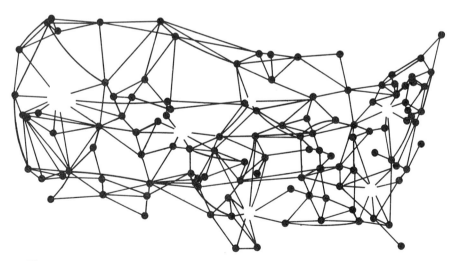

Figure 8.2   114-Node Sample Network.

tigate the effects of alternate routing on network performance. We consider both local and global adaptive routing and observe that these routing algorithms can sometimes improve and at other times deteriorate network performance.

Let us set aside adaptive routing for a moment and consider what happens to a network using fixed, sequential alternate routing and crankback controls under damage and congestion. Fig. 8.1 and 8.2 present the undamaged and damaged networks, respectively, under consideration. The damaged network is obtained from the undamaged network by removing 5% of the undamaged network's nodes (both of these networks are described in more detail later). The results are summarized in Table 8.1. For these results, the engineered load is 2,000 Erlangs. Under damage, overload results from the loss of nodes and/or links which forces the remaining facilities to handle all the offered load to the network.

By looking at this table clearly, even a network with relatively powerful controls, alternate routing, and crankback is quite susceptible to the small amounts of overload and damage considered. Therefore, we consider adaptive routing to further improve the performance of networks under stress.

Table 8.1 Network Performance Under Damage and Congestion

Performance Measure	No Damage 2,000 Erlangs	Damaged 1,500 Erlangs	Damaged 2,000 Erlangs
Call completion probability	0.92	0.76	0.50
Average call set-up time	51.4	112.4	170.1
Average number of probes per call	5.9	11.1	15.2
Percent of successful probes per call	95	68	52
Average number of links per call	4.8	6.2	6.6
Average number of crankbacks per call	0.00	0.52	1.45

In the remainder of this chapter we present a study carried out to investigate the use of adaptive routing techniques to improve the performance of networks under stress. In the remainder of this section, Section 1, we will describe the characteristics of the networks considered and the various adaptive routing techniques to be employed. Section 2 presents the model used for the investigation and makes clear the assumptions used in the study. The computer simulation developed to implement the model is covered in Section 3. Section 4 presents the results of our study.

1.1 Network Characteristics

The sample network used for this study, Fig. 8.1, consists of 120 nodes and 299 bi-directional links. All nodes and links have fixed (finite) capacities and no queuing. The network has been sized to carry 2,000 Erlangs with a call completion probability of at least 0.92. The sizing was carried out, for a network in which all nodes are in operation, using fixed alternate routing. An average node in this sized network has a capacity of 116 simultaneous calls.

The damaged network, Fig. 8.2, is a subset of the above (sized) network consisting of 114 nodes and 299 bi-directional links. The assumption being made for this case is that six nodes in the network are not operational. This represents a 5% node loss with an average call-carrying capacity of 110 calls for the undamaged nodes, and an average call-carrying capacity of 221 calls for the damaged nodes. Thus, it can be seen that the damaged nodes were "key" call-carrying nodes. No links are assumed to be damaged.

At each node in the network there is a routing table for each destination that may be reached from that node. The routing table consists of an ordered list of nodes that are possible subsequent nodes on the call's path to its destination. The node at the top of the list represents the best choice, i.e., on the shortest path to the destination. The last entry on the list represents the worst choice. Each node selects the first entry on its routing table and attempts a probe to that node. If the probe fails, the node tries a probe to the second element on its routing table, continuing in this way until either a successful probe is made or the node runs out of choices on its routing table. This is the sequential alternate routing capability. If a successful probe is found, the call progresses to the next node and from there selects a new node in a similar fashion. The call thus progresses in this way, node by node, until it reaches its destination or is blocked. If a node is unable to successfully probe any of the nodes on its routing table, and this is not the origination node for the call, the node then returns control of the call to the previous node on the call's path, crankback. The previous node then continues probing its routing table, starting at the position following that of the node from which the call cranked back. When the node originating a call probes unsuccessfully all the entries on its routing table, the call is blocked. In general, a call is blocked if any of the following events take place:

1. The call path exceeds a specified length.
2. The call setup time exceeds a specified limit.
3. The origin node is busy or damaged at the time of call origination.
4. A call cranks back all the way to the originating node and this node's routing list is exhausted.
5. The destination node is busy or damaged.

With this description of network characteristics we now discuss the local and global adaptive routing algorithms to be investigated.

1.2 Adaptive Routing

This section discuses the various adaptive routing algorithms, both local and global, considered in our network study. These algorithms will then be used on the 120 node and 114 node networks already described.

The local adaptive routing algorithms considered are described in [3] and reviewed here only briefly. We begin by assuming that we are at node n_1 trying to set up a call to node n_2. Each node has a routing table, indexed by destination, for each possible destination (i.e., for each node in the reachable set from n_1). The routing table at n_1 for destination n_2 is $\{p_1, p_2, \cdots, p_k\}$. Suppose now that probes to nodes p_1 and p_2 fail and that the third probe (recall we are using sequential routing tables) to node p_3 is successful. Here the local adaptive routing algorithm Success To Top (STT) will reorder the routing table by moving the successful probe to the top of the list, resulting in the new routing table $\{p_3, p_1, p_2, p_4, \ldots, p_k\}$. On the other hand, the algorithm Failure To Bottom (FTB) will change the routing table by moving the failed probes to the bottom of the list, resulting in the new routing table $\{p_3, \cdots, p_k, p_1, p_2\}$. Note that STT moved the successful probe to the top and displaced the failed probes by one position, leaving $p_4, \ldots, p_k$, the untried nodes, in the same position as before. FTB also places the successful probe at the top of the list; however, it moves the two failed probes all the way to the bottom of the list and thus effectively moves up $p_4, \ldots, p_k$ two positions without any knowledge of whether these are indeed better nodes to probe than p_1 and p_2. As it will be seen later, this is not a very good algorithm for the applications considered here.

The third local adaptive routing algorithm used is called Average Number of Links (ANL). This algorithm associates an ordered list of weights $\{w_1, w_2, \ldots, w_k\}$ with each routing table satisfying $w_i \leqslant w_{i+1}$. The weight w_i corresponds to routing table entry p_i. In our example, ANL would increase the value of w_i, $i = 1,2$ by some function of w_i and decrease w_3 by some function, depending on the number of links the call required to reach its destination through node p_3. The w_i, $i = 4, \ldots, k$ remain unchanged. Once ANL is done changing the w_i's, it reorders the routing table so that the weights are once again arranged in increasing order. Depending on how much the w_i's changed, it may or may not be necessary to actually reorder the routing table.

The performance of the various local adaptive routing algorithms will be compared with No Learning (NL), which is the case when sequential alternate routing is used with no table reordering.

A problem with having only decentralized local control in the networks considered here is that information diffusion is relatively slow and thus the quality of the local feedback is poor. Without a method for disseminating information the adaptive routing algorithms at each node in the network can learn from other nodes only indirectly; this can lead to a detrimental effect on the network's performance. When the adaptive routing algorithms operate in a dynamically changing network, the information diffusion rate may be too small to be of any use to the network in globally adapting to the change. A coherence length naturally arises and can be defined as the distance at which information becomes unreliable, relative to the frequency of change in the network state [4]. Clearly, the coherence length becomes longer as the information diffusion rate increases. If the coherence length is not greater than the diameter of the network's graph, then an incompatibility between adaptive routing algorithms in different parts of the network exists, and global instability will occur, resulting in poorer overall performance, at times even poorer than a network with no learning. Therefore, we also consider global adaptive routing algorithms that use global network status information to reorder routing tables. It is not clear that one would want to use local algorithms if global algorithms are implemented. However, when we use both local and global techniques, it is important to properly define the way in which they interact with each other to make them compatible.

The global adaptive routing technique considered uses global status information dissemination. This information dissemination is carried out with a superimposed data network whose sole function is to disseminate network status information. Network status information will change more rapidly the more detailed it is. Since the data network will have some bit rate limitation, there will be a limit to the detail of the state information on which the global control can act. The structure and data rate of the data network will determine what characteristics of the circuit-switched network can be globally controlled at all. If the data network is packet-switched, then there are two basic modes to route information: point-to-point and broadcast. In the point-to-point mode a packet might be transmitted from origin to destination according to some routing tables, while in the broadcast mode the packet would be replicated to nearest neighbors in a flooding fashion. For this study we will use flooding and a data transmission rate for the packet network of 300 baud. The important thing to understand, given this limitation and the need to update information fast enough to properly reflect network dynamics, is how much information each node can pass. It is assumed that each node will pass the same type and amount of information and thus the information passed by each node results in packets of equal length.

The information dissemination scheme for the packet-switched network is carried out by having all nodes, at a specified time, send their node status information to a given subset of this set of nodes. If this subset of nodes consists of a single node, this node will then process all the information and, in some fashion, give the rest of the nodes the information they need to properly reorder their routing tables. The assumption being made is that such global information reflects the current network status more accurately than that gathered by each node based on their respective probing experiences, and thus that the routing tables can be reordered in a way that would increase network throughput and call completion probability and decrease call set-up time. The subset of nodes receiving the status information may consist of more than a single node, each gathering the information relevant to their neighbors (to whom this information is not disseminated). These nodes can then do some preliminary processing on the data received from their "subject" nodes and disseminate it among themselves. More processing would then be done to incorporate all the global data, and finally the appropriate data would be transmitted to the rest of the nodes (the subject nodes) so that they may reorder their routing tables based on global information.

The method used for the present study was to let the subset of nodes participating in direct global information sharing consist of the entire network with the information being passed in a somewhat nonredundant flooding fashion, the reason for this being that we would like to maintain a completely distributed network. As soon as a few nodes are designated as somehow "special," the network's performance is then directly dependent on these nodes remaining operational. If such is not the case, serious network performance degradation can result.

Given that all nodes will take part in global information dissemination and processing, it is necessary to decide when, how often, and what information each node must disseminate and how to process, at each node, all the gathered information. The approach taken in this study is to disseminate information on a session basis. That is, we determine the time, denoted D, it takes for all nodes to disseminate their information to all other nodes and set the session time (S) to be no less than D. Every S units of time, all nodes disseminate their status data. S is set to a value needed to properly reflect network dynamics, but always satisfying

$$S \geqslant D$$

This approach assures that all queues are cleared before new information is disseminated. As a result, we are assured that congestion in the packet-switched network will not be a problem and that messages disseminated at a given time arrive at all nodes before new data are disseminated. Within a session, whether all nodes start disseminating information at the same time or in some sequence does not seem to have a big effect on the calculated time D, as long as the data network nodes have buffers able to handle the resulting queues.

If the needed session time S turns out to be less than the calculated time D, it will be necessary to reduce the amount of information being disseminated per node or increase the data transmission rate, until D and S are equal.

The data each node disseminates in our study consist of the probability (based on the node's experience) of successfully probing each of its nearest neighbor nodes. For the networks described, the average node contains approximately four to five nearest neighbors. The packet length of information needed to disseminate was set at 50 bits. By a packet-switched simulation it was determined that D was less than 40 seconds. S was set at 120 seconds, which is believed to be appropriate for the network dynamics being modeled.

At the end of the session, once each node has received all probabilities p_{ij}, denoting the probability of successfully probing node j from node i, it uses the last n (for some prespecified number n) set of p_{ij}'s to calculate a running average $\bar{p}_{ij}$ given by

$$\bar{p}_{ij} = \left[\sum_{k=1}^{n} p_{ij}^{(k)} \right] / n$$

where $p_{ij}^{(k)}$, $k = 2,..., n$, is the value of p_{ij} obtained $(k-1)$ sessions prior to the current session whose probability is $p_{ij}^{(1)}$. The $\bar{p}_{ij}$'s are then used to obtain the new routing tables using a k-shortest path algorithm. In this algorithm, the distance between adjacent nodes i and j is given by

$$d_{ij} = l_{ij} - \alpha \ln \bar{p}_{ij}$$

where l_{ij} is the distance (in number of links) between nodes i and j, which for our nearest neighbor case takes on a value of one. α was also set to one. Thus as the probability of successfully probing node j from node i tends to zero, d_{ij} will tend to infinity. On the other hand, as the probability tends to one, then so does the distance between adjacent nodes i and j. The best routing tables are obtained when calculating $\bar{p}_{ij}$ for large values of n.

2. MODEL OF THE NETWORK

We have thus far described the network characteristics, network controls, and adaptive routing algorithms to be used in our study of network performance under stress. We now present a description of the model developed for the investigation. As already mentioned, the type of networks considered have fixed capacity nodes with no queuing of calls at the nodes. Thus when a call arrives at a node, it either accesses the node or it probes an alternate node, if there are any. Links between nodes also have finite capacity. The performance of such networks is to be evaluated subject to varying levels of congestion and node outages using various local and global adaptive routing techniques. A secondary objective of the model is to be used for sizing the network, in other words, to arrive at the trunk group (link) and switching system (node) capacities required to carry a given traffic load while meeting a specified grade of service. Network sizing is directly dependent on the chosen network operating environment and adaptive routing scheme.

In Section 2.1 we describe the base model [5] which is a detailed model of the envisioned network and its operating environment. Owing to its size and complexity, it is not possible to use such an all-encompassing model for our study of network performance. Thus, Section 2.2 describes a lumped model which is a simplified version of the base model yet appropriate for our studies. The lumped model was used to derive the computer simulation described in Section 3. From a comparison of the base and lumped models it will be possible to understand the various assumptions made in the computer simulation and thus to put into proper perspective the performance results obtained using the simulation.

2.1 Base Model

Our starting point, or base model, for the network consists of the transmission and switching facilities of a telecommunications network (which can consist of properly interconnected public and private telecommunications networks). A subset of all nodes (i.e., switching systems) is designated during the design stage as the nodes to make up the network for study. Routing at the network nodes is based on call destination and consists of a sequential routing table (for each destination reachable from this node) composed of adjacent (meaning, logically connected) nodes to which the call can be routed on its way to the destination. These routing tables may be reordered via some adaptive routing algorithm based on network status and call progress feedback. An important feature assumed for this network is crankback which allows a node to return control of a call to the previous node, on the call's path, when unable to successfully probe any of the entries on its routing table. After a successful probe, the probing node passes information to the next node on the path describing (among other things) all the nodes currently forming part of the call. This information may then be used to avoid the formation of call loops in the network by requiring that a node being probed is not already part of the call. The success of a probe depends on the availability (both in terms of congestion and damage) of the corresponding nodes and links. Once a node is successfully probed and node information has been passed, the new node takes control of the call by selecting (from its routing table) and probing the next node on the call's path to the destination. When a call reaches its destination node, a voice path through the network has been established and remains in effect for the duration of the call. When the call completes (or when a probing call is blocked), the facilities being used by that call are released. The time it takes for all facilities to be released is dependent on the number of facilities in use for that particular call. There is a maximum number of nodes that can be used to set up a call before the transmission is impaired. Calls that need more than this maximum number of nodes are blocked.

For this network, traffic is assumed to be highly time-varying in terms of pattern, call interarrival rates, and call holding times. Furthermore, throughout the network's operation it is expected that the availability of nodes and links will change depending on network congestion, network repair, and further network outages.

2.2 Lumped Model

The lumped model, used to derive the computer simulation, models the network. Its main components are network nodes, logical connections, network traffic, routing algorithm, and routing tables. Thus, in the lumped model we ignore the possibility of non-network traffic competing for needed resources. This is not an unreasonable assumption if it is expected that network traffic is to be protected from public traffic via some scheme (e.g., trunk reservation). Furthermore, we have lumped together network transmission and switching facilities interconnecting pairs of nodes and treat them as network logical connections. A pictorial representation of the base model and lumped model networks is given in Fig. 8.3. Fig. 8.3a represents a portion of the base model network composed of switching systems, network nodes, and trunk groups operating under both network and non-network traffic. The corresponding lumped model network, Fig. 8.3b, consists of network nodes and logical connections operating under network traffic only. In this figure we also show the routing tables at each node; these tables are included in the base model (although not shown in Fig. 8.3a). The node-to-node probing time is randomly (and uniformly) chosen from a given range of values. This is in contrast to the base model in which the probing time depends on the actual number of switching systems making up the logical connection. When a call completes (or is blocked), all facilities (i.e., nodes and logical connections) used by the call are released instantaneously.

Descriptive variables for each of the network components are shown in Table 8.2. The anticycling parameter in the above table selects the particular anticycling rule to be used by the model. We consider three possibilities; the first of these is weak anticycling which requires that a node being probed cannot already form part of the call. Strong anticycling has the same constraint as weak anticycling with the added restriction that the chosen node cannot be a node from which the call previously cranked

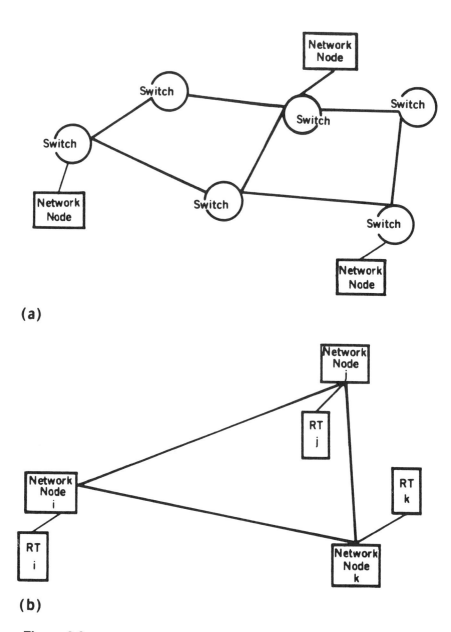

Figure 8.3   (a) Base Model Network. (b) Lumped Model Network.

Table 8.2 Network Component Descriptive Variables

Component/Variable	Description
Nodes:	
capacity	positive number denoting node capacity, a capacity of zero denotes a damaged node
time-out	amount of time a node waits for acknowledgment that a probe succeeded
Routing Tables:	
node list	list of nodes to be probed by a given node in setting up a call
Trunk Groups:	
capacity	positive number denoting connection capacity, zero capacity denotes damaged connection
tmepr	time required for probe to travel node-to-node connection
Routing Algorithm:	
anticycling	anticycling mode used
tblptr	specifies next probe on routing table after crankback
learn	local adaptive routing: 1 = yes, 0 = no
globtime	session time
globinc	number of session between table reconstructions
Network Traffic:	
gent	call interarrival rate
choldt	call holding time
traffic	traffic distribution used

back. Strong anticycling avoids making probes that have a low probability of success, thereby (it is hoped) reducing call setup time. The last case is no anticycling in which any node chosen is acceptable.

When a call cranks back from, say, node i to node j (i.e., node i was a dead end), node j must choose from its routing table the next node to probe. In doing this it can choose the node at the top of the list. Since choices from the routing table are made sequentially, the top entry must have been probed unsuccessfully, and it may even be node i. A second approach for node j is to choose the entry on the routing table below node i. This would assure that the chosen probe has not been (recently) attempted for this call. Either of these choices may be acceptable and which one of these is in use is controlled by *tblptr*.

Global adaptive routing is modeled by assuming a data network used to distribute network status information to all nodes in the network. This is done by allowing each node to disseminate, at the same time over the data network, information describing the nodes' operational status. The session time (controlled by *globtime*) denotes how often each node can (and should) disseminate its status information. The parameter *globinc* determines how often (in terms of sessions) the routing tables are to be reconstructed using the global information. A list of model parameters appears in Table 8.3. The parameter *maxtan* controls the maximum number of nodes that make up a call. This parameter controls the amount of network facilities that can be devoted to a single call and should be low enough to guarantee acceptable transmission. The *tsetup* parameter controls the amount of time that the network should spend in setting up a single call. The parameters *alpha* and *beta* are discussed later. The network component interactions can be described as follows:

1. The network traffic model determines the call interarrival rate, geographic traffic distribution and call holding time. When a call arrives at the network, the origination and destination nodes are chosen randomly according to some specified distribution.
2. When a call arrives at a node, the node searches its routing table for the call's destination and selects the next node to be probed subject to anticycling constraints. A probe is then sent out on the logical connection between the two nodes. If the logical connection is not busy or damaged, the probe reaches the new node. If this node is not busy or damaged, the probe is successful and information is sent identifying all nodes currently taking part in the call. This new node then looks for the next node to probe. The call continues in this fashion, from node

Table 8.3   Model Parameters

Parameter	Description
maxtan	maximum number of tandem nodes allowed
tsetup	maximum amount of time a call is in probing status
alpha	sizing parameter
beta	sizing parameter

to node, until either the destination node is reached or the call is blocked.

3. When the destination node is reached, the call is connected, and the nodes and logical connections are utilized for the duration of the call (*choldt*). At this point, if local adaptive routing is used, all nodes on the call's path will reorder their tables based on the successful route for this call.

4. Once the call completes, all nodes and logical connections are instantaneously released to handle other calls.

5. A call will be blocked if:
   - Either of the parameters *maxtan* or *tsetup* are exceeded.
   - The origin node is busy or damaged at the time of call origination.
   - A call cranks back all the way to the origin node and the routing table of this node is exhausted.
   - The destination node is busy or damaged.

6. Throughout the simulation run, global adaptive routing (if in use) obtains network status information from nodes and periodically (as controlled by *globinc*) reconstructs the routing tables of all nodes based on the gathered data and some global adaptive routing algorithm.

The traffic model used is a uniform traffic model (all nodes have equal probability of being chosen as the origination or destination of any given call). The call interarrival rate and call holding time are randomly (and uniformly) chosen from a range of numbers specified by the user. The lumped model also assumes that repair or damage of facilities does not occur during the times of network operation being modeled.

3. COMPUTER SIMULATION

A computer simulation was written in the C language to implement the network model described. This section presents a brief description of the simulation along with a description of its main applications.

3.1 An Overview of the Simulation Process

The main input to the simulation consists of files defining the network component descriptive variables (i.e., network topology) described in Table 8.2. The rest of the input consists of the model and simulation parameters, Tables 8.3 and 8.4. Damaged links or nodes can be modeled by initializing their capacities to zero. Calls are generated based on a

Table 8.4 Simulation Parameters

Parameter	Description
iroute	Number of elements in the routing table
nnode	Number of nodes simulated
noc	Maximum number of calls to be generated during a simulation run
maxtime	Maximum simulation run time
subtotals	Number of subtotal (intermediate) statistics to be printed out
nost	Maximum number of simultaneous calls allowed in the network
trace	1 = Print out paths of individual calls 0 = No trace print out
statime	Interval over which steady-state statistics are gathered
sttime	Statistical sampling time for graphical output
restart	0 = Simulation is run from empty initial state 1 = Simulation is restarted from earlier run
seed	Random number generator seed

predetermined traffic model with a given call generation (interarrival) rate. The simulation ends when either a specified amount of simulated time has elapsed or a specified upper bound on the number of calls generated has been reached.

Given a network consisting of several nodes and interconnecting links, a brief example of how the simulation handles a call setup is now given. We begin by trying to set up a call from node i to node j. In order for the call to progress a port must be available at node i; otherwise, the call is blocked. If node i has a free port, the routing table is checked to find the next node on the path from i to j. Let us say it's node k. First we must see if a trunk is available from i to k. If a trunk is not available, the probe is blocked and the routing table is checked for the second choice of routes from i to j. If a trunk is available, a probe can be attempted at node k only if the node is not already on the path from i to j (to avoid cycling). If the probe is successful (a port is available at node k), information is passed to the node describing, among other things, the identity of all nodes currently taking part in this call. The information transfer time τ is given by

$$\tau = \alpha_1 + \alpha_2 \nu$$

where ν denotes the number of nodes (excluding node k) already taking part in the call, and α_1 and α_2 are appropriate transmission rate constants. The probing process continues from node to node until the destination is reached or the call is blocked. If at any point in the call, the routing table for the current node to the destination has been exhausted (every node on the table has been probed and the probes failed), control of the call is transferred to the previous node in the call path (called crankback). Calls are blocked under the conditions specified in the last section.

3.2   Network Sizing and Performance

The simulation can be used to size network facilities (i.e., to obtain the necessary capacities at each node and link interconnecting each node pair for a given traffic load and desired blocking level), as well as to gather network performance measures. In the first case the user defines a certain level of blocking that will be acceptable from a network user's point of view. Then by assigning an infinite capacity to both types of facilities the simulation estimates the facility capacities at which the desired blocking occurs. This is done by assigning an initial value to the expected sizes and sampling the facility sizes at uniform increments of time. At each time increment, if the current expected size is less than the current size needed at that facility (for zero blocking), the expected size is increased by a parameter α. If the current expected size is greater than the current size needed at that facility (for zero blocking), the expected size is decreased by a given parameter β. Here α and β are chosen to satisfy

$$\frac{\alpha}{\alpha + \beta} = 1 - \rho$$

where ρ is the desired blocking probability. Once the network has been sized, one can proceed to evaluate the network's performance by looking at the various network performance graphs and statistics. For either of these types of computer simulation runs, a restart capability is available. In other words, a user can start the simulation from an initial empty state in the network or restart it from a previous run, thus having the capability to start the simulation run from any desired network state.

3.3 Routing

The simulation is written in a modular fashion. This simplifies the implementation of different adaptive routing algorithms. All the local adaptive routing algorithms so far considered, all of which reorder the routing tables when a call successfully connects to its destination, are available to the simulation. Also available is global adaptive routing which uses global network status information to reconstruct routing tables. Any of these algorithms may be used in a particular simulation run by appropriately selecting a parameter when submitting the simulation run.

4. STEADY-STATE ANALYSIS

Using the above simulation, we now present results on the performance of networks under stress, congestion and node outages, using the various local and global adaptive routing algorithms discussed.

4.1 Local Adaptive Routing Performance

This section presents results concerning local adaptive routing. We consider three scenarios of network operation:

• Congested network with all nodes operational.
• Uncongested network with node outage.
• Congested network with node outage.

For this study, a congested network means a network operating under a traffic load of approximately 2,000 Erlangs, whereas an uncongested network has a traffic load of approximately 1,500 Erlangs. In either case, the topology of the network with node damage is given by the 114-node network of Fig. 8.2, and the network topology when all nodes are operational is given by the 120-node topology in Fig. 8.1. As previously mentioned, the 120-node network has been sized (using NL) for a 2,000 Erlang traffic load, so that by congestion, in this case, we mean that the network is operating at capacity.

Table 8.5 presents the performance statistics for the 120-node network with a 2,000 Erlang offered load. Note that this is the situation for which

Table 8.5 120-Node Network Without Global Learning
Rate of 0.5 Seconds and Call Interarrival

Performance Measure	NL	ANL	STT	FTB
Call completion probability	0.92	0.90	0.74	0.58
Average call set-up time	51.4	53.7	83.2	98.9
Average number of probes per call	5.9	6.1	8.4	11.6
Average % of successful probes per call	95	95	85	82
Average number of links per call	4.8	4.9	6.4	7.3
Average number of crankbacks per call	0.00	0.01	0.10	0.35

the network has been engineered, that is, the topology and traffic load are as expected. We can see from the table that indeed the network performs well for the case of NL; however, network performance deteriorates when using ANL, STT, or FTB with the latter performing noticeably worse than NL. Fig. 8.4 presents the carried load (solid curves) and the probing load (dashed curves) as a function of time for each of the four algorithms. It is seen here that the highest carried load (i.e., number of connected calls in the network) is achieved using NL.

It is indeed a bit disconcerting to have these local adaptive routing algorithms, which we hoped would improve network performance, actually degrade performance. It should, however, be remembered that the simulation was started with the routing tables in optimal order (i.e., the initial routing tables are ordered by path length to the destination, with the node leading to a shortest path length placed at the top of the table); thus, any reordering of the tables will lead to poorer performance. The key is to find the algorithm that least degrades performance for a network already operating optimally and improves performance when operating under congestion and/or node damage. It will be seen that ANL falls into this category.

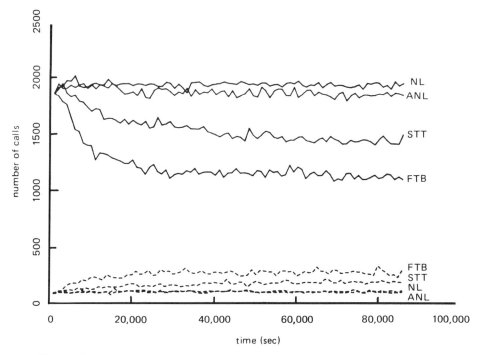

Figure 8.4   Carried Load (solid) and Probing Load (dashed) Curves for the
120-Node Network with 2000 Erlangs.

The following set of performance statistics, Table 8.6, presents results
for the damaged 114-node network with an offered load of approximately
1,500 Erlangs. In this situation we see that local adaptive routing algorithms improve network performance (actually only ANL and STT, with
STT improving performance only slightly). The corresponding carried and
probing load curves are shown in Fig. 8.5. Note that in this case the
highest carried load and lowest probing load are achieved using ANL.

We now have seen a case in which local adaptive routing deteriorates
network performance and one in which it improves performance. The last
set of performance measures in this section, Table 8.7, are for the damaged
network with congestion (2,000 Erlangs). From Fig. 8.6 it can be seen
that the carried load is approximately the same whether NL or ANL is
used.

Table 8.6 114-Node Network Without Global Learning
and Call Interarrival Rate of 0.75 Seconds

Performance Measure	NL	ANL	STT	FTB
Call completion probability	0.76	0.85	0.77	0.70
Average call set-up time	112.4	74.8	91.5	97.2
Average number of probes per call	11.1	8.1	9.6	11.1
Average % of successful probes per call	68	88	86	86
Average number of links per call	6.2	5.9	7.0	7.4
Average number of crankbacks per call	0.52	0.14	0.23	0.31

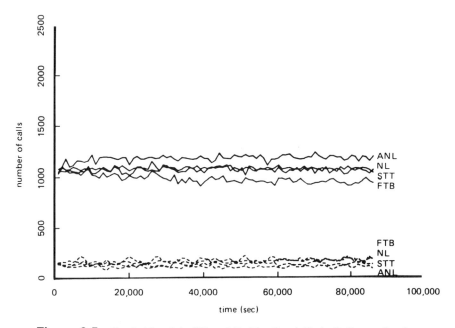

Figure 8.5 Carried Load (solid) and Probing Load (dashed) Curves for the 114-Node Network with 1500 Erlangs.

Table 8.7 114-Node Network Without Global Learning and Call Interarrival Rate of 0.50 Seconds

Performance Measure	NL	ANL	STT	FTB
Average call completion probability	0.50	0.53	0.45	0.34
Average call set-up time	170.1	138.0	166.8	172.1
Average number of probes per call	15.2	12.6	14.3	16.7
Average % of successful probes per call	52	59	59	58
Average number of links per call	6.6	6.6	7.6	7.8
Average number of crankbacks per call	1.45	1.08	1.41	1.72

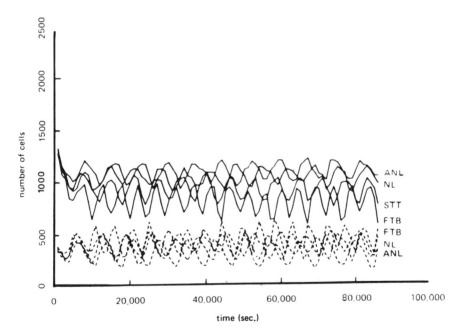

Figure 8.6 Carried Load (solid) and Probing Load (dashed) Curves for the 114-Node Network with 2000 Erlangs.

In this case we see that there is no real improvement in performance to be gained by using local adaptive routing. The small improvement under ANL is attributed to an imbalance in the levels of congestion and node outages to which the network was subjected. The improvement can be eliminated by a slight variation in the offered load.

We summarize these observations as follows:

• For networks with congestion and all nodes operational - local adaptive routing deteriorates network performance.
• For networks with little congestion and node damage - local adaptive routing improves network performance.
• For networks with some congestion and node damage - there is no noticeable improvement over NL when using a local adaptive routing algorithm.

Of course these results are also dependent on network connectivity. We have found similar results for a number of networks with node degrees in the 4-6 range.

Intuitively these observations can be explained as follows. In the first case, congestion with all nodes operational (i.e., 120-node topology), one can deduce that since the tables are already optimally ordered and since blocking due to congestion is a temporary problem (congestion is not high enough to keep nodes constantly blocked), it is best not to change the initial tables. The penalty of such a change is a longer path length, leading to a longer setup time, and more congestion due to the (perhaps) wasteful use of facilities. On the other hand, when there is little congestion but with node damage (i.e., the 114-node network), it is appropriate to move the routing choices corresponding to damaged nodes toward the bottom of the tables since, as assumed in our model, these nodes will never again be operational during the simulation run.

The case of node damage and congestion seems to imply that the problems of congestion and damage are somewhat independent of each other and that each approach (learning or no learning) takes care of only one of these two problems. Thus, in the situation of damage and congestion both local adaptive routing and no learning result in basically the same network performance. Performance is nevertheless poor for this case, and something better is needed.

The following section presents preliminary results using global adaptive routing. The idea is to improve performance for the case of networks operating under both node damage and congestion by improving the quality of the feedback information used by the adaptive routing algorithms.

4.2   Global Adaptive Routing Performance

This section presents preliminary global adaptive routing results using the techniques previously discussed. The improvement in network performance using global adaptive routing can be seen by comparing this section's results, Table 8.8, with those of Table 8.7.

A comparison of the two tables, column by column, shows that an improvement in performance can be detected for all the performance measures considered whether or not global adaptive routing is used in conjunction with a local adaptive routing algorithm. The corresponding carried and probing load curves appear in Fig. 8.7. The improvement in performance is not as good as might be expected. It must be remembered, however, that there are yet many parameters, associated with global adaptive routing, whose effect on the network remains to be determined. Furthermore, the choice of information used for dissemination is not necessarily optimal. It was shown in [1] that when using "perfect learning" (i.e., routing tables are reconstructed using actual network topology information instead of probabilistic information), global adaptive routing has a marked effect on network performance. Thus, it is necessary to properly tune parameters and consider different status data for dissemination to correctly ascertain the worth of global adaptive routing in such networks.

Table 8.8   114-Node Network With Global Learning and Call Interarrival Rate of 0.5 Seconds

Performance Measure	NL	ANL	STT	FTB
Call completion probability	0.54	0.54	0.53	0.53
Average call set-up time	141.8	136.9	139.3	138.9
Average number of probes per call	12.2	12.0	12.1	12.2
Average % of successful probes per call	60	61	60	61
Average number of links per call	6.6	6.6	6.6	6.6
Average number of crankbacks per call	1.08	1.07	1.09	1.10

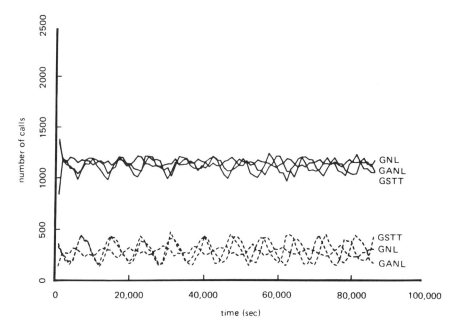

Figure 8.7 Carried Load (solid) and Probing Load (dashed) Curves for the 114-Node Network with 2000 Erlangs Using Global Adaptive Routing.

REFERENCES

[1] Garzia, M. R. and Lockhart, C. M. - "On the Performance of Congested Networks with Distributed Intelligence," *Modeling IEEE TC on Simulation* (20), pp. 19-25, July 1984.

[2] Garzia, M. R. and Lockhart, C. M. - "On the Design and Performance of Nonhierarchical Networks with Distributed Control (U)," Proceedings of the 1984 IEEE Military Communications Conference, October 1984.

[3] Meketon, M. S. and Topkis, D. M. - "Adaptive Routing for Damaged Networks," Proceedings of the IEEE Military Communications Conference, October 1985.

[4] Narendra, K. S. and Thathachar, M. A. L. - "Learning Automata - A Survey," *IEEE Transactions on Systems, Man and Cybernetics* SMC-4 (4), July 1974.

[5] Zeigler, B. P. - *Theory of Modeling and Simulation*. Wiley, 1976.

9

Modeling Network Dynamics

Mario R. Garzia

AT&T Bell Laboratories
Columbus, Ohio

Clayton M. Lockhart

AT&T Bell Laboratories
Holmdel, New Jersey

1. COMPARTMENTAL ANALYSIS OF NETWORKS

In this chapter we develop compartmental models to investigate the dynamic behavior of networks under stress, i.e., damage and congestion. We then use these models to investigate various network management control schemes. We consider two compartmental models for circuit-switched networks. The state variables chosen for the models reflect the basic driving factor of network dynamics, the use of network resources. Details about the microstate of a network, that is, the exact configuration of calls on the network, are considered to be unimportant compared to the amount of node and link capacity used by the calls. As a result, the model will be spatially homogeneous. The model's describing variables (i.e., state variables) consist of global system measures, such as probing and carried load. We do not model the delay and blocking due to either call register queue overflow or queue length control mechanisms. To do so would take us out of the compartmental model paradigm and into discrete event dynamical systems. The differential/difference equation approach would no longer apply because variables would evolve from past variables evaluated at different times. The system variables would not evolve in a uniform discrete time step owing to the delays caused by call register queues.

The processors and links associated with the signaling network will be assumed to be highly efficient so that signaling congestion will not be modeled except that a finite time delay, incorporated implicitly in a discrete time delay factor that also includes a switch central processor delay factor, is accounted for in the process of either reaching or not reaching a subsequent node. Our model only considers congestion on the trunk groups. A simple customer retrial model is included as part of the basic dynamical equations. Blocking caused by damage is modeled by uniform blocking factors applied to trunk groups and switches, with the implicit assumption of random damage which is the only possible form of damage given the assumption of homogeneity in the model.

Traffic is assumed to be distributed evenly at all times throughout the network so that each trunk group blocking probability is the same. Traffic parcels offered to each trunk group in the network are assumed to be statistically independent and to follow a standard Poisson arrival process. Customers are assumed to retry with a certain probability and in a certain average time if they are blocked.

2. NETWORK MODELS

In this section we develop the two basic compartmental models used for our study. In later sections, these models are expanded to incorporate the various network management control schemes to be investigated. We consider highly connected networks, so that a call path between any two nodes on the network will consist of a one-link or two-link path. For our compartmental models we consider six classes of calls (six classes of compartments): calls reattempting access to the network, denoted by P_0; calls on their first link, on a one-link call path, denoted by P_{11}; calls on their first link of a two-link path, denoted by P_{21}; calls on their second link of a two-link path, denoted by P_{22}; calls that are connected through a one-link path, denoted by C_1; and calls that are connected through a two-link path, denoted by C_2. The difference between the two models developed is in the total number of compartments (or call states) required. For the first model, the call populations P_{21} and P_{22} are broken up into different populations, one for each of the alternate paths possible. The second model, which we call the lumped model, does not make such a distinction. Thus, the total number of state variables for the first model is $4 + 2r$, where r is the number of two-link routes on a routing table. A second model is described requiring only six state variables and is referred to as a lumped model, as it is an averaged version of the more detailed (first) model. For large r the lumped model is computationally faster. Of course, by lumping together the information for each ith path into a single variable some resolution is lost and therefore this model is not as accurate. However, for the cases studied, the two models provide comparable results.

2.1   Circuit-Switched Network Model

The macroscopic variables used throughout the network models are listed as follows:

C_1 : Average number of connected calls using one link
C_2 : Average number of connected calls using two links
P_{11} : Average number of calls probing the destination directly from the origin
P_{21}^i : Average number of calls probing the first link of the ith two-link path $(i = 1,...,r)$
P_{22}^i : Average number of calls probing the second link of the ith two-link path $(i = 1,...,r)$
P_0 : Average number of failed probes that will reattempt

We also define the following parameters:

N : Number of switching nodes in the network
d : Number of trunk groups emanating from each node
r : Number of two-link routes on each node's routing table $(r > 0)$
n : Number of trunks per trunk group
ν : Inverse call holding time
μ : Inverse probe reattempt time
λ_0 : Bare call arrival rate (time-varying)
τ : Inter-node probing time
p_B : Probability of one or more free trunks in the trunk group
p_R : Probability of call reattempt
p_S : Switch survival probability
p_T : Trunk group survival probability

Any failure probability q (perhaps with subscript) will refer to the quantity $1 - p$ (perhaps with subscript), Where p is a success probability. Difference equations for these state variables can be written down by inspection of the state diagram shown in Fig. 9.1. The state diagram presents the various compartments, input from the environment and output to the environment, and the flow rates (transitions) between states. Δ will always refer to the forward time difference operator with time step τ. The

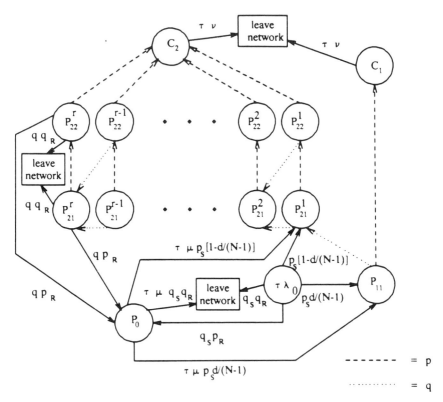

Figure 9.1   State Transition Diagram for the Circuit-Switched Network Model.

difference equations for the global variables in which the time step is set at τ are as follows:

$$\Delta P_{11} = -P_{11} + \tau\lambda_1 \tag{9.1}$$

$$\Delta P_{21}^i = -P_{21}^i + q(1-\delta_{1i})(P_{21}^{i-1} + P_{22}^{i-1}) + \delta_{1i}(qP_{11} + \tau\lambda_2) \tag{9.2}$$

$$\Delta P_{22}^i = -P_{22}^i + pP_{21}^i \tag{9.3}$$

$$\Delta P_0 = -\tau\mu(p_S + q_Sq_R)P_0 + p_Rq(P_{21}^r + P_{22}^r) + \tau\lambda_0q_Sp_R \tag{9.4}$$

$$\Delta C_1 = -\tau\nu C_1 + pP_{11} \tag{9.5}$$

$$\Delta C_2 = -\tau\nu C_2 + pP_{22} \tag{9.6}$$

where $i = 1,\ldots,r$, and δ_{1i} is given by

$$\delta_{1i} = \begin{cases} 1 & \text{if } i = 1 \\ 0 & \text{if } i \neq 1 \end{cases}$$

and where the following convenient expressions have been defined:

$$P_{21} = \sum_{i=1}^{r} P_{21}^i \tag{9.7}$$

$$P_{22} = \sum_{i=1}^{r} P_{22}^i \tag{9.8}$$

$$\lambda = \lambda_0 + \mu P_0 \tag{9.9}$$

$$\lambda_1 = \frac{p_s d\lambda}{N-1} \tag{9.10}$$

$$\lambda_2 = \frac{p_S(N-d-1)\lambda}{N-1} \tag{9.11}$$

$$p = p_B p_S p_T \tag{9.12}$$

λ is the "renormalized" call attempt rate which includes the effect of reattempts (P_0) due to network blocking. Define quantities L_e and L_i by

$$L_e = \frac{C_1 \nu + 2C_2 \nu + P_{22}\tau^{-1}}{C_1 + 2C_2 + P_{22}} \tag{9.13}$$

$$L_i = \frac{2p_T p_S}{Nd\tau}(P_{11} + P_{21} + P_{22}) \tag{9.14}$$

L_i is the rate at which free trunks are being probed while L_e is the average rate at which trunks in use are freed. The ratio $E = L_i/L_e$ is the "effective offered load" on the average network trunk group. If we assume Poisson arrivals, the trunk group blocking is given by (see the Appendix for a derivation)

$$q_B = 1 - p_B = B(n,E) \tag{9.15}$$

where $B(n,E)$ is the Erlang B formula [14] for E Erlangs offered to n fully available trunks in a trunk group. For integer n, $B(n,E)$ is given by

$$B(n,E) = \frac{\dfrac{E^n}{n!}}{\sum_{i=0}^{n} \dfrac{E^i}{i!}} \tag{9.16}$$

and can be calculated by a recursive formula [13]. The difference equations are nonlinear because the Erlang B formula depends on the global variables.

2.1.1 Steady-State Behavior

If we assume various fixed values of the parameters, the steady-state behavior of the network was obtained by running the difference equation into steady state for a range of offered loads, damage factors, and customer retrial behavior. The fixed parameters have the following values which are typical of the AT&T DNHR network: $N = 100$ nodes; $d = 80$ trunk groups; $r = 5$ routes; $n = 120$ trunks; $\nu = 1/300$ Hz; $\mu = 1/60$ Hz; $\tau = 0.5$ sec. A reattempt probability of $p_R = 0.8$ and the median reattempt time of $1/\mu = 60$ sec is taken from [11]. The variable parameters have the following ranges: $p_S = p_T \in \{0.2, 0.3, 0.4, 0.5, 0.6, 0.7, 0.8, 0.9, 1.0,\}$ and $p_R \in \{0.0, 0.8.\}$ To compute steady-state behavior λ_0 is varied quasi-statically and various network performance measures are averaged over sufficiently long time intervals. Five performance measures monitored are the carried load, $C = C_1 + C_2$, the link blocking probability, q, the end-to-end blocking probability, $B = 1 - p(P_{11} + P_{22})/\lambda\tau$, the average number of links $L = 1 + C_2/(C_1 + C_2)$, and the call set-up time, T, which is calculated by the formula:

$$
T = \tau \left[\frac{\dfrac{d}{N-1} + \left[1 - \dfrac{pd}{N-1} \right] \left[\dfrac{1+p}{p^2} - (1 - p^2)^r \left[\dfrac{1+p}{p^2} + \dfrac{r(1+2p)}{1+p} \right] \right]}{1 - \left[1 - \dfrac{pd}{N-1} \right](1 - p^2)^r} \right]
$$

$$(9.17)$$

The performance measures plotted against the bare offered rate λ_0, for the case of reattempts and the nine damage levels, are shown in Fig. 9.2 to 9.6. Note the hysteresis effects in the cases of low to moderate damage, implying that there are regions of bistability. At higher levels of damage the bistability (although not the inefficiency of call routing due to congestion) disappears because a damaged network does not allow as many calls on it, and the calls which do manage to get on the network find enough capacity to complete to surviving destinations. The retrial of blocked calls causes a more pronounced hysteresis effect than the case of no retrial, although the basic bistable behavior is similar in this situation. We do not show the no-retrial curves since they are analogous to those in Fig. 9.2

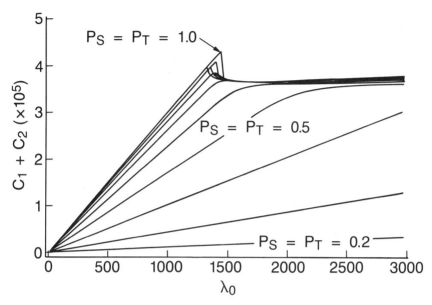

Figure 9.2   Carried Load Hysteresis for Different Damage Levels.

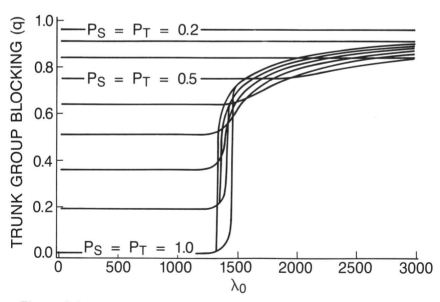

Figure 9.3   Trunk Group Blocking Hysteresis for Different Damage Levels.

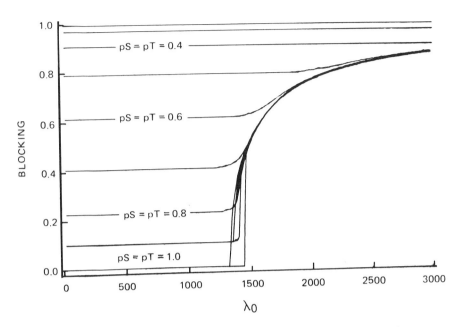

Figure 9.4 End-To-End Blocking Hysteresis for Different Damage Levels.

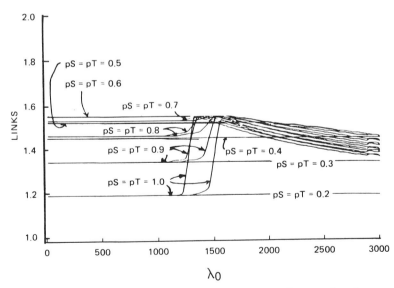

Figure 9.5 Number of Links Hysteresis for Different Damage Levels.

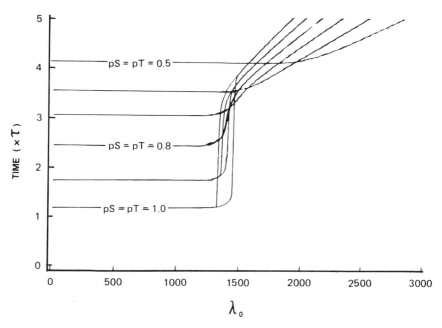

Figure 9.6   Set-Up Time Hysteresis for Different Damage Levels.

through 9.6. From these figures one concludes that the nonhierarchical networks considered here basically operate in one of two modes: efficient or inefficient. Fig. 9.5 shows that the inefficient state is characterized by the population of two-link calls saturating the network resources, while the efficient state is dominated by one-link calls. Note that, once beyond the region of bistability, the carried load does not drop off as λ_0 is increased, because single link calls become increasingly dominant at this level of congestion. The carried load would drop off if switching delays were included in our model [5,12].

The behaviors of the three basic Grades-Of-Service (GOS) measures, B (end-to-end blocking), L (number of links), and T (call set-up time), vary with the number of two-link alternate routes, r. Alternate routing can both help and hurt the network, depending on the level of congestion. In steady state these GOS measures can be expressed as functions of p, the link transmission probability. For simplicity of expression, we assume that

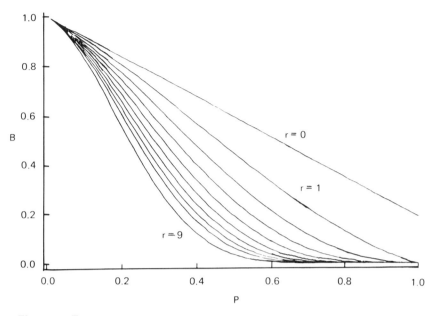

Figure 9.7 End-To-End Blocking Under Steady State.

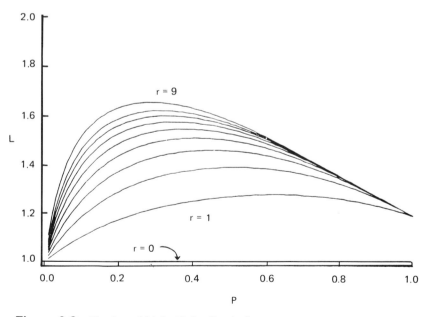

Figure 9.8 Number of Links Under Steady State.

$p_S = p_T = 1$ and $p_R = 0$ (no retrial). The steady-state expression for T is given above by Equation (9.17). B and L are given by

$$B = \left[1 - \frac{pd}{N-1} \right] (1-p^2)^r \qquad (9.18)$$

$$L = 1 + \frac{\left[1 - \dfrac{pd}{N-1} \right] \left[1 - (1-p^2)^r \right]}{1 - \left[1 - \dfrac{pd}{N-1} \right] (1-p^2)^r} \qquad (9.19)$$

Fig. 9.7 through 9.9 show the behaviors of B, L and T, respectively, as functions of p for values of $r = 0,1,...,10$. One might infer from these figures that a higher value of r results in a better GOS; however, Fig. 9.10 demonstrates that this is not the case. Fig. 9.10 shows a family of hysteresis curves for C (carried load) for $r = 1,2,...,10$. Note that as r

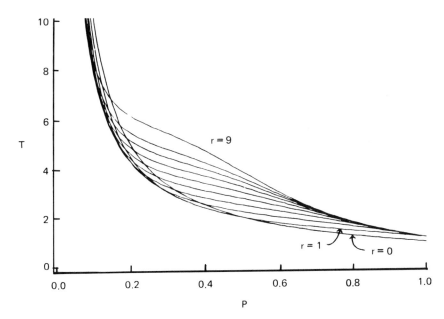

Figure 9.9   Set-Up Time Under Steady State.

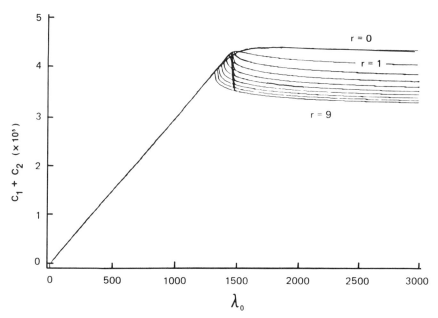

Figure 9.10   Carried Load, No Reattempts.

increases, the hysteresis begins earlier (at smaller λ_0) and is worse (more area in the hysteresis cycle). This result has implications for the behavior of an expansive network management control which tries to increase the number of alternate routes when blocking occurs. If the network is running at $r = 1$ and some slight blocking is occurring (B increases slightly), one might try to reduce end-to-end blocking by increasing the number of alternate routes available (on the basis of Fig. 9.7) and, for instance, set $r = 5$. Then, depending on the value of λ_0 ($1300 \leqslant \lambda_0 \leqslant 1450$ for the case of $r = 5$), the system can be pushed into a bistable region, and a transition to the inefficient state can occur with large enough environmental fluctuations. Without any restrictive or selective traffic controls, expansive routing techniques can actually be harmful to network performance.

2.1.2   Time Behavior

We have seen the steady-state behavior of the nonhierarchical network under overload and damage. It is interesting to investigate the network's behavior through time under various stress situations. In a later section on network management controls the stressed network is required to perform

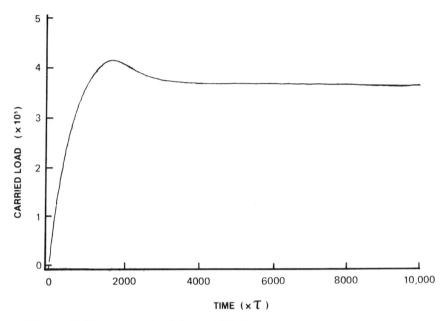

Figure 9.11   Carried Load, $\lambda_0 = 1500$, Reattempts.

up to maximum capability by proper application of controls. The network with no controls yields baseline comparison cases against which to test network controls. All the cases in this section on time behavior will assume reattempts.

The simplest stress situation is constant overload. Fig. 9.11 shows the carried load behavior over a period of 10,000 τ, where $\lambda_0 = 1500$ (1500 calls attempt to enter the network every second). There is an overshoot and then the network settles to an inefficient state, since $\lambda_0 = 1500$ represents about a 13% overload. Fig. 9.12 shows how trunk blocking behaves in the same period. For an uncontrolled network things become interesting only when the environmental stress is changing. Therefore, a variable traffic and damage situation was placed on the network as shown in Fig. 9.13. The damage comes in three successive waves (at 4000 τ, 5000 τ, and 6000 τ) with the last wave causing a significant reduction in offered load. Before the first wave, and through the second wave, of damage the traffic is increasing up to a maximum of about three times the engineered load. This stress model is to be representative of crisis traffic during an attack on

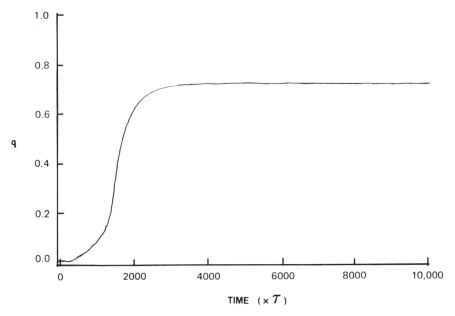

Figure 9.12 Trunk Group Blocking, $\lambda_0 = 1500$, Reattempts.

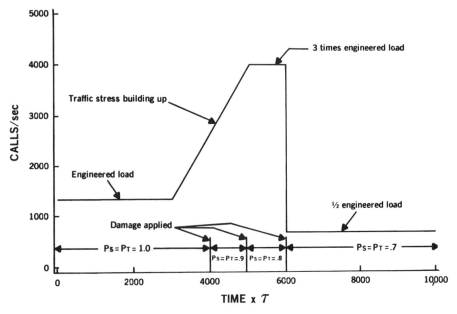

Figure 9.13 Traffic and Damage Stress Scenario.

250

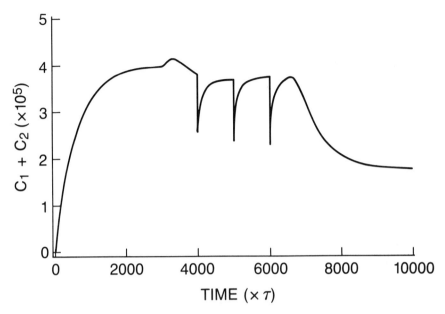

Figure 9.14 Carried Load in the Stress Scenario, Reattempts.

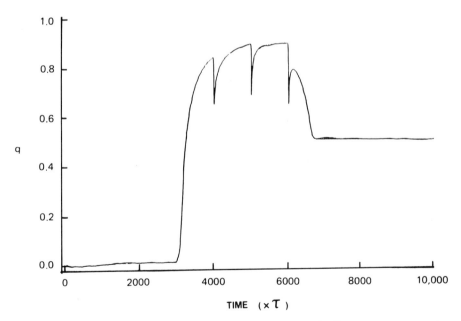

Figure 9.15 Trunk Group Blocking in the Stress Scenario, Reattempts.

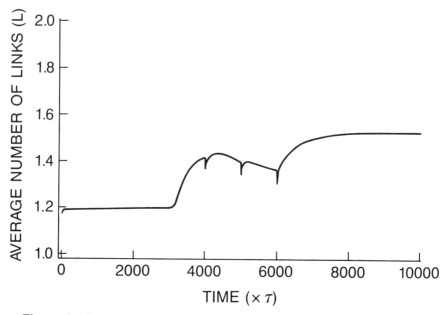

Figure 9.16   Average Number of Links in the Stress Scenario, Reattempts.

the network. Fig. 9.14 shows the carried load under this stress scenario. The traffic "hump" which appears after the last wave of damage is a call surge caused by reattempts, while the sharp downward jumps show the loss of calls due to damage. Fig. 9.15 shows trunk blocking with similar downward jumps resulting from congestion relief due to damage. Fig. 9.16 shows that the average number of links tends to increase during the extreme stress period, demonstrating network inefficiency associated with the lack of controls during network stress. In the period after the last wave of damage the number of links GOS goes even higher; however, the blocking is lower because there is enough capacity in the network to handle calls during this time.

2.2   Lumped Circuit-Switched Network Model

We now develop the lumped model version of the model discussed above. The new model has only six state variables leading to a reduced dimensionality of the state space. Just as in the previous sections, this model is of the general form

$$\dot{x}(t) = A(x,t)\, x(t) + B(t)\, u(t) \tag{9.20a}$$

$$y(t) = C(x,t)\, x(t) \tag{9.20b}$$

where $x \in \mathbb{R}^6$, $u \in \mathbb{R}$, and $y \in \mathbb{R}^X$. The matrices A, B, and C are time-varying and of appropriate dimension, with A a function of x. Using the terminology of the last section, we now define the state vector x in (9.20) as follows

$$x = \begin{pmatrix} x_1 \\ x_2 \\ x_3 \\ x_4 \\ x_5 \\ x_6 \end{pmatrix} \triangleq \begin{pmatrix} P_0 \\ P_{11} \\ P_{21} \\ P_{22} \\ C_1 \\ C_2 \end{pmatrix} \tag{9.21}$$

The input variable $u(t)$ is simply the bare arrival rate $\lambda_0(t)$.

As mentioned, this model has reduced state-space dimension compared to the model of the last section whose state-space dimension was a function of r, the number of two-link paths per call. Of course, there is a loss in detail as a result of lumping together these state variables.

Using the definitions from the previous section and looking at the state diagram of Fig. 9.17, we see that the matrix A is given by

$$A \triangleq \begin{pmatrix}
-\mu(1 - p_R q_S) & 0 & p_R q \zeta / \tau & (p_R q \zeta / \tau) & 0 & 0 \\
p_S \mu d / (N-1) & -1/\tau & 0 & 0 & 0 & 0 \\
p_S \mu [1 - d/(N-1)] & q/\tau & -(q\zeta + p)/\tau & (q/\tau)(1 - \zeta) & 0 & 0 \\
0 & 0 & p/\tau & -1/\tau & 0 & 0 \\
0 & p/\tau & 0 & 0 & -\nu & 0 \\
0 & 0 & 0 & p/\tau & 0 & -\nu
\end{pmatrix}$$

where we define

$$\zeta = \frac{p^2 (1 - p^2)^{r-1}}{[1 - (1 - p^2)^r]}$$

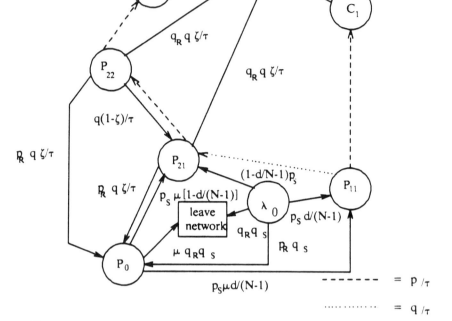

Figure 9.17   State Transition Diagram for the Lumped Circuit-Switched
Network Model.

The vector $B \in \mathbb{R}^6$ is defined as

$$B \triangleq \begin{bmatrix} p_R q_S \\ \left[\dfrac{d}{N-1}\right] p_S \\ \left[1 - \dfrac{d}{N-1}\right] p_S \\ 0 \\ 0 \\ 0 \end{bmatrix} \tag{9.22}$$

The basic difference between this and the previous model is that in this version we do not explicitly model the various possible alternate routes. Instead we lump all the $P^i{}_{21}$ and $P^i{}_{22}$ into two variables P_{21} and P_{22}, respectively. Therefore, the difference is in the representation of these two variables and their interactions with the other state variables. The remaining state variables and their interactions, P_0, P_{11}, C_1, C_2, remain unchanged, except of course for their interactions with P_{21} and P_{22}. In contrast with the previous model, all calls do not leave state P_{21} in time τ, since some fail on their first few attempts until they reach a successful probe to a via switch. States P_{21} and P_{22} contain a heterogeneous population. In P_{21} the population is made up of r classes of calls. For $0 \leqslant i < r$, the ith class consists of those calls that have failed the last i attempts to connect; the number of calls in this class is given by

$$(1 - p^2)^i \ P_{21}$$

This scaling is then used to collapse Equations (9.1) through (9.6) into the lumped model equations (9.20) through (9.22).

There are various output functions $y(t)$ that are appropriate for our study. One output function of interest can be defined as

$$C(x,t) \ = \ C \triangleq (0,0,0,0,1,1) \tag{9.23}$$

which represents system carried load. This and other output maps will be used later on. As before, we see that the nonlinearity of the A matrix comes from q_B which is defined by Equation (9.15).

2.3 Steady-State Behavior

We now present the steady-state results obtained using the lumped version of the network model. The parameters settings were the same as those used for the previous model, namely, $r = 5$, $n = 120$, $\nu = 0.0033$, $\mu = 0.0167$, $\tau = 0.5$, and $p_R = 0.8$. Fig. 9.18 presents the carried load curves for the various levels of damage, $p_S = p_T \in \{0.2, 0.3, 0.4, 0.5, 0.6, 0.7, 0.8, 0.9, 1.0\}$. As can be seen from this figure, the results closely resemble those of the earlier model, including the regions of bistability leading to the inefficient operating states. This will be the case for the rest of the results presented, thus showing that the lumped version of the model is appropriate for representing various qualitative and quantitative features of the network. Fig. 9.19 and 9.20 represent the blocking, q, and trunk group blocking, q_B, probabilities, respectively. Fig. 9.21 shows the average number of links for the various damage levels.

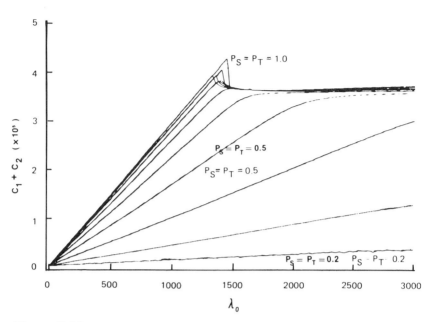

Figure 9.18 Lumped Model Carried Load.

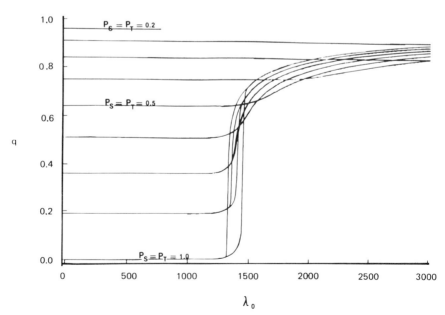

Figure 9.19 Lumped Model Trunk Group Blocking Probability.

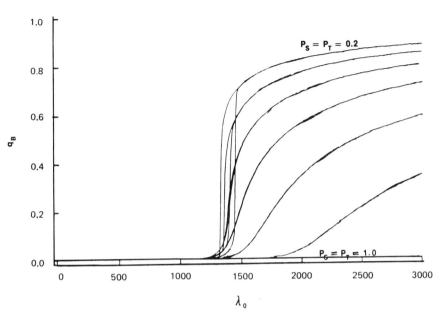

Figure 9.20 Lumped Model Trunk Group Blocking Probability (traffic component q_B only).

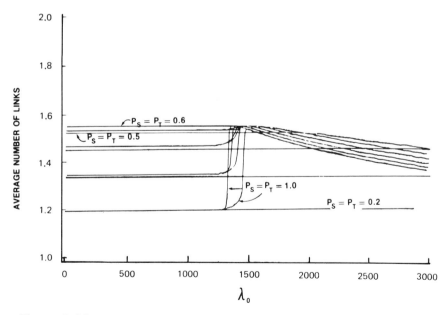

Figure 9.21 Lumped Model Average Number of Links.

Curves showing the time behavior of the system are presented later, where they are compared with access denial control curves.

3. NETWORK MANAGEMENT CONTROLS

3.1 General Discussion

Since an uncontrolled network under stress can run inefficiently, it is of interest to consider the performance of network management controls. The purpose of these controls is to insure that the network will continue to operate in an efficient, revenue-producing mode. In a nonhierarchical network, network congestion can cause a runaway effect if it is not checked early [9]. Saturation of a certain resource, such as trunk capacity, causes calls to increase their use of alternate routing, thereby using additional network resources per call and causing further congestion in other resources, such as switching. Under severe overload the resource to congest first is generally the trunking, followed by the switching resources (call registers). The result is that an uncontrolled nonhierarchical network can exhibit bistable behavior, efficient and inefficient modes of operation [7,15,17]. We shall observe this behavior and show that certain network management controls are able to do away with this bistability. A review of network management techniques for networks with stored program controlled switching can be found in [6,12,16]. Network controls can be categorized as expansive or restrictive. An example of an expansive control might be to allow for increased alternate routing when blocking becomes too high for certain routes. If portions of the network are congested while others are relatively free from traffic congestion, then expanded alternate routing can make use of idle capacity and be beneficial to call completion. By the nature of our (homogeneous) compartmental network model, traffic inhomogeneities cannot occur (or be accounted for); therefore, only restrictive controls will be considered. Restrictive (protective) controls reduce the routing flexibility and decrease the completion probability of certain kinds of calls so that the network becomes less congested and hence will run in the efficient state. Too much routing flexibility precipitates a saturation effect in the network and an inefficient traffic handling capability. Restrictive controls should be adaptive, and depending on the level of congestion in the network, the controls should block varying amounts of traffic. If congestion has a rapidly time-varying behavior, the controls should react quickly to check congestion before the situation becomes worse or to back off quickly when the congestion is relieved. The restrictive controls should not be overly rigid so that the maximum possible carried load (revenue) is achieved for the given offered load.

As mentioned earlier, there are basically three kinds of network congestion in a network: switching, trunking, and signaling congestion. Switch delay and congestion are not modeled here because to do so would take us out of the compartmental model paradigm and into discrete event dynamical systems. Signaling congestion is not modeled, mainly because the transmission rates are increasing as the signaling network becomes increasingly modernized. A finite time delay, incorporated implicitly in the inter-node probing time τ, takes into account the signaling process of either reaching or not reaching a subsequent node. Only trunk congestion is modeled.

In the next section we define the network management controls to be considered for our study. Then in Sections 3.3 through 3.5 we enhance the network models to incorporate these controls and use them to investigate the resulting network performance.

3.2   Controls and Activation Mechanisms

Two kinds of restrictive controls and the related activation parameters will be considered. Generally, network management controls are non-smooth because they are activated by applying threshold criteria to success and failure statistics collected by switching systems. If the failure rate goes above a first threshold, a control will be activated. The control will be deactivated once the failure rate falls below a second threshold (not necessarily equal to the first threshold). Normally, different thresholds (hysteresis controls) are used to prevent "chatter" from occurring; however, we will not discuss hysteresis controls here and will consider single-threshold models only.

The first threshold parameter we consider employs measurements of the ineffective network attempt (INA) rate [11], defined by the number of probes which ultimately fail, given that they succeeded in being forwarded to the next switch, divided by the total number of network attempts (NA) successfully forwarded to the next switch. The INA rate statistic (which we will denote by $R = $ INA/NA) is generally calculated with a 5 minute window [11], which happens to be close to the average long-distance call holding time. The INA rate usually includes the call attempts that fail to complete for any of three reasons: the destination line number does not answer (DA), the destination is busy (BY), or equipment blockage and failure (EB&F). Since we only model customer behavior insofar as a retrial model and do not model the DA and BY behavior, the INA rate will be limited to failed attempts from the EB&F category. The second threshold parameter is simply the trunk group blocking probability, q, which is an internal property of the network not perceived by customers. The third threshold parameter is the number of free trunks in a trunk group.

The restrictive network management controls considered in this study adaptively control network parameters such as the input offered rate λ, and the blocking of certain calls relative to others. If the offered load is such that a measured quantity, such as the ineffective network attempt rate or the trunk group blocking, exceeds a threshold, then the offered rate, λ, could be reduced by blocking at the originating switch, a control called "access denial." A trunk group control called "trunk reservation" has an effect on the network when the number of trunks occupied in a trunk group rises above a certain threshold. Only the preferred calls are allowed to use the remaining trunks once the threshold has been reached, enabling those calls to receive a higher completion probability at the expense of other, less preferred calls. We will consider two kinds of preferred calls. When the network is congested, preferred calls are defined as those which use fewer network resources. First routed calls can be given precedence and lower blocking, thereby causing two-link calls to experience a higher blocking factor and reducing congestion due to alternate routing. Another kind of preference involves splitting the offered load λ_0 into two components, Class I and Class II, where the Class I population is a small fraction of the total. Trunks would be reserved for the Class I calls, improving their throughput under overload. Such a trunk reservation feature is useful for high-priority users such as the Government customers who need high completion during crises, or perhaps for high revenue-producing customers.

3.3 Access Denial Model

To model the restrictive control of access denial some changes to the difference equations are necessary. An extra difference equation for the access probability p_A is needed to reflect the adaptive feedback control of access denial. p_A is the probability of accessing the first switch before routing of the call begins. If we use the variable definitions of the previous sections and the state variable flow diagram shown in Fig. 9.22, the difference equations are as follows:

$$\Delta P_{11} = -P_{11} + \tau\lambda_1 \qquad (9.24)$$

$$\Delta P_{21}^i = -P_{21}^i + q(P_{21}^{i-1} + P_{22}^{i-1})$$
$$+ \delta_{1i}(qP_{11} + \tau\lambda_2) \qquad (9.25)$$

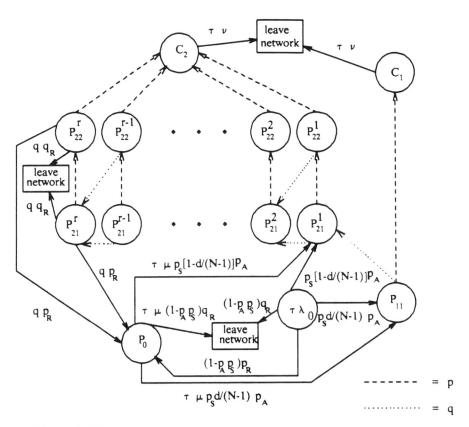

Figure 9.22   Access Denial Model.

$$\Delta P_{22}^i = -P_{22}^i + pP_{21}^i \tag{9.26}$$

$$\Delta P_0 = -\tau\mu[p_S p_A + (1 - p_S p_A)\,q_R]P_0$$
$$+ p_R q(P_{21}^r + P_{22}^r)$$
$$+ \tau\lambda_0(1 - p_S p_A)\,p_R \tag{9.27}$$

$$\Delta C_1 = -\tau\nu C_1 + pP_{11} \tag{9.28}$$

$$\Delta C_2 = -\tau\nu C_2 + pP_{22} \tag{9.29}$$

$$\Delta p_A = -\epsilon[\,sgn(x - x_c) + \theta(x_c - x)\,\delta_{1,p_A}] \tag{9.30}$$

where x refers to either q or R, depending upon which measured parameter the access denial is to be triggered by. The parameters ϵ and x_c, respectively, refer to an access blocking adjustment factor and a critical threshold for a measured network parameter, respectively, and

$$\theta(x) = \begin{cases} 1 & \text{if } x > 0 \\ 0 & \text{if } x \leqslant 0 \end{cases}$$

Consistent with Equation (9.30) the access denial mechanism operates in the following manner. If the measured quantity x is greater than x_c, then p_A is decremented. If the measured quantity x is less than x_c, then p_A is incremented, unless p_A is already unity in which case no change to p_A occurs. λ_1 and λ_2 are defined by:

$$\lambda_1 = \frac{p_S p_A d\lambda}{N-1} \tag{9.31}$$

$$\lambda_2 = \frac{p_S p_A (N-d-1)\lambda}{N-1} \tag{9.32}$$

$$\lambda = \lambda_0 + \mu P_0 \tag{9.33}$$

The INA rate, R, in the specific context of the nonhierarchical switched networks considered here, is simply the time average (nominal 5 minute window) of the trunk blocking, q. The number of calls that succeed in being forwarded to the next (via) switch and are subsequently blocked (ineffective network attempts) is given by qP_{22}, while the number of calls that have been forwarded to the next switch (network attempts) is P_{22}. As a result, $R = INA/NA = <qP_{22}/P_{22}> = <q>$, with the brackets representing a time average. This simplification of the expression for R is the result of a two-link routing restriction.

3.3.1 Steady-State Behavior

There is no difference, in steady state, between access control triggering on either q or R, because, in our case, R is simply a time average of q. Fig. 9.23 and 9.24 show the steady-state plots of the carried load and link blocking probability, respectively. Reattempts are included in the system to compare with previous results. Each of the curves used the critical value of $q_c = 0.05$ and $\epsilon = 0.01$, which turn out to be nearly optimal, as we will see. Fig. 9.23 and 9.24 are to be compared directly with Fig. 9.2 and 9.3, which pertain to the uncontrolled network. Note that with the access denial control the dramatic hysteresis effects are gone in all cases of damage and that the carried load is essentially maximized.

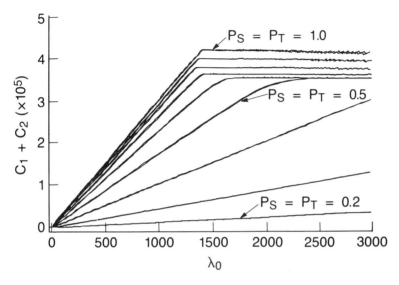

Figure 9.23   Carried Load, $q_c = 0.05$, $\epsilon = .01$, Reattempts.

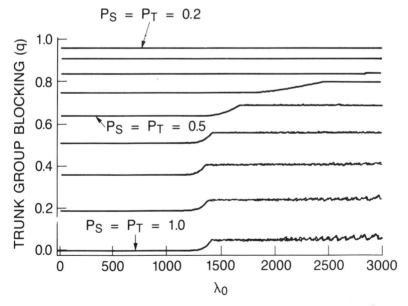

Figure 9.24   Trunk Group Blocking, $q_c = 0.05$, $\epsilon = .01$, Reattempts.

For the sake of optimal network control in steady state, it is necessary to know how the parameters q_c and ϵ cause the carried load to change. The steady-state carried load was monitored for ranges of both q_c and ϵ, resulting in the family of curves shown in Fig. 9.25. The values $q_c = 0.13$, $\epsilon = 0.01$ yield approximately the highest carried load, although the optimum is fairly broad. Clearly, the access blocking adjustment parameter, ϵ, should be small for the best *steady-state* result. However, one would expect that, in a highly transient situation, ϵ would have to be larger so that the feedback mechanism could react more quickly.

3.3.2 Time Behavior

As in the case of no controls, we look at the constant overload situation, in which $\lambda_0 = 1500$, for access denial triggered by either q or R. For either trigger $x_c = 0.05$ and $\epsilon = 0.01$. As usual, we consider the reattempt case only ($p_R = 0.8$). The time averaging window for R is set at 300τ, which

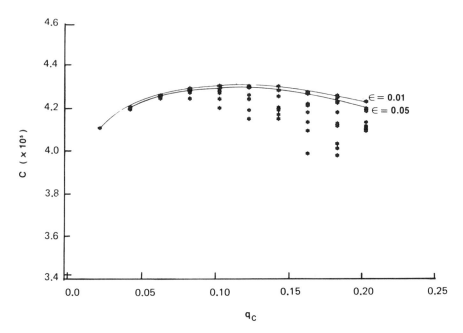

Figure 9.25 Carried Load Varying Control Settings.

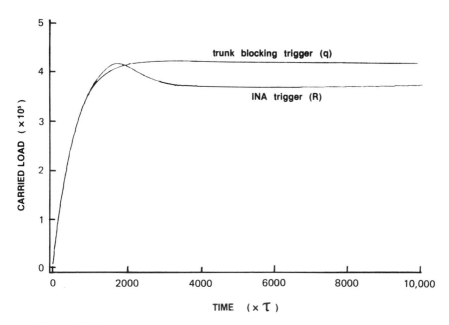

Figure 9.26 Carried Load/Reattempts, $\lambda_0 = 1500$.

ıs 150 sec, and hence is shorter than the normal integration window by a factor of two. Fig. 9.26 shows the effect of the access denial control on the carried load. Note that the control triggered by the INA ratio does poorly, mainly because it is so slow in reacting to the overload situation. Fig. 9.27 shows the effect of the access denial control on the link blocking with a similar difference in behavior. Note the rather peculiar time behavior of the link blocking for the case of the q trigger: first there is a smooth transient, then a random phase, then regular oscillation, and then complete quiescence. This rather unpredictable, fast time-scale behavior for q is characteristic of the q-triggered access denial mechanism. The nonlinear feedback through p_A causes self-generated stochastic behavior. In comparing Fig. 9.28 with Fig. 9.11 (no controls and overload) it is clear that q-triggered access denial helps the network in this simple stress situation.

Turning to the stress scenario (Fig. 9.14), we can now investigate the effect of the q-triggered access denial control. Fig. 9.28 shows the time

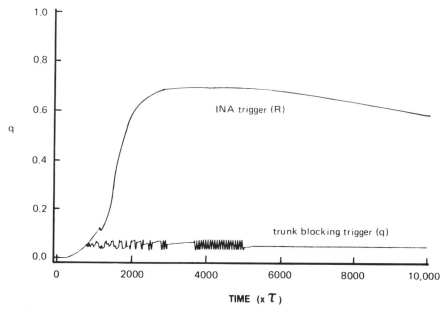

Figure 9.27 Trunk Group Blocking/Reattempts, $\lambda_0 = 1500$.

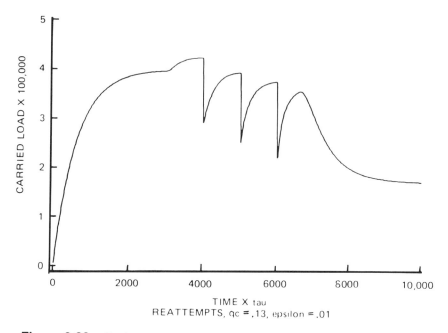

Figure 9.28 Carried Load in the Stress Scenario, $q_c = .13$, $\epsilon = .01$, Reattempts.

behavior of the carried load with the access denial control ($q_c = 0.13$, $\epsilon = 0.01$) and with call reattempts. This figure should be compared with Fig. 9.14, which shows the uncontrolled network behavior under the same stress scenario. Note that access denial does not increase the carried load much over the uncontrolled case, and that it actually decreased the carried load part of the time. The conclusion is that access denial has a marginal utility in protecting the network, mainly because access denial blindly discriminates against all traffic, not just those calls that use the most network resources per call. The situation is quite different for the trunk reservation control discussed in the next section.

3.4 Trunk Reservation Model for First-Routed Calls

As already explained, the network management scheme of trunk reservation for first-routed calls results in preferential treatment for those probing states which represent first-routed calls. Explicitly, first-routed calls are those states P_{11} and the fraction of the states P_{21}^1, P_{22}^1 which do not overflow from P_{11}. Preferential treatment for the preferred probes is effected by reserving access to a specified remaining fraction of a trunk group for the first-routed calls. That is, if the average trunk group load goes beyond a certain value $m < n$ trunks, alternate-routed calls will be blocked. This discrimination between call probes causes a difference in blocking as felt by first-routed and alternate-routed calls. For the sake of computational simplicity, a slightly different definition of "preferred states" will be discussed here. The preferred states will be defined as P_{11}, P_{21}^1, and P_{22}^1. Calls that overflow P_{11} to the first two-link route are still given preferential treatment.

With the usual variable definitions, the total rate at which free trunks in the average trunk group are attempted on is

$$L_i = \frac{2p_T p_S}{Nd\tau}(P_{11} + P_{21} + P_{22}) \tag{9.34}$$

while the rate at which first-routed calls attempt on free trunks in the average trunk group is given by

$$L_i^1 = \frac{2p_T p_S}{Nd\tau}(P_{11} + P_{21}^1 + P_{22}^1) \tag{9.35}$$

The fraction of first-routed calls attempting on free trunks is $f = L_i^1/L_i$, while the effective offered load is still $E = L_i/L_e$, where L_e is given by

$$L_e = \frac{C_1 \nu + 2C_2 \nu + P_{22} \tau^{-1}}{C_1 + 2C_2 + P_{22}} \tag{9.36}$$

Define a normalization factor X by

$$X = \sum_{i=0}^{m} \frac{E^i}{i!} + \sum_{i=m+1}^{n} \frac{E^i}{i!} f^{i-m} \tag{9.37}$$

The trunk group completion probabilities p_B^1, p_B^2, respectively, for first-routed and alternate-routed calls, respectively, are then given by [1]

$$p_B^1 = p_B^2 + X^{-1} \sum_{i=m}^{n-1} \frac{E^i}{i!} f^{i-m} \tag{9.38}$$

$$p_B^2 = X^{-1} \sum_{i=0}^{m-1} \frac{E^i}{i!} \tag{9.39}$$

The quantities p_B^1 and p_B^2 can be calculated using recursive formulae [10]. Let $p_1 = p_B^1 p_S p_T$, $p_2 = p_B^2 p_S p_T$, and $\phi = p_1 - p_2$. The difference equations can then be modified to accommodate trunk reservation for preferred calls, by inspection of Fig. 9.29.

$$\Delta P_{11} = -P_{11} + \tau \lambda_1 \tag{9.40}$$

$$\Delta P_{21}^i = -P_{21}^i + (q_2 - \delta_{2i}\phi)(P_{21}^{i-1} + P_{22}^{i-1}) \\ + \delta_{1i}(q_1 P_{11} + \tau \lambda_2) \tag{9.41}$$

$$\Delta P_{22}^i = -P_{22}^i + p_2 P_{21}^i + \delta_{1i}\phi P_{21}^1 \tag{9.42}$$

$$\Delta P_0 = -\tau \mu (p_S + q_S q_R) P_0 \\ + p_R (q_2 - \delta_{r1}\phi)(P_{21}^r + P_{22}^r) \\ + \tau \lambda_0 q_S p_R \tag{9.43}$$

$$\Delta C_1 = -\tau \nu C_1 + p_1 P_{11} \tag{9.44}$$

$$\Delta C_2 = -\tau \nu C_2 + p_2 P_{22} + \phi P_{22}^1 \tag{9.45}$$

3.4.1   Steady-State Behavior

The steady-state behavior of the network with trunk reservation control eliminates hysteresis. GOS steady-state measures will not be shown, as they are similar to those shown in the case of access denial. The optimal value of m, the reservation threshold, can be determined in steady state for various traffic and damage conditions. Fig. 9.30 shows the steady-state carried load as a function of m, for various network stress situations: 13%

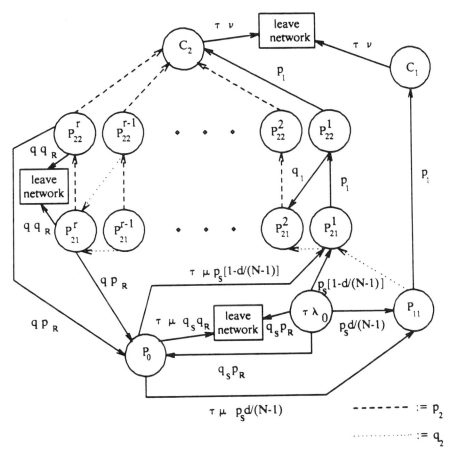

Figure 9.29   Trunk Reservation Model.

overload (with and without retrial), 100% overload (with retrial), 100% overload and damage ($p_S = p_T = 0.8$) (with retrial), and 5% overload (with retrial). Most of the curves have three different regimes of behavior. For low values of m the curves are flat because p_2 is approximately zero, which implies that all calls complete using the preferred probing states P_{11}, P_{12}^1, P_{22}^1. The peaked region represents a higher throughput brought about by an appropriate balance between the use of the preferred probing states and further alternate-routing states. The carried load falls dramati-

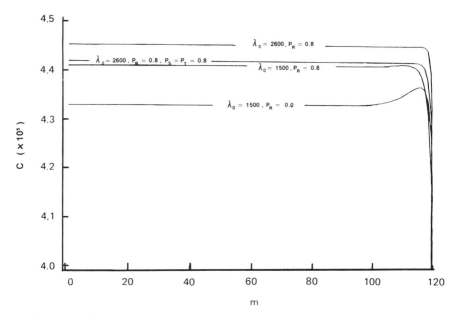

Figure 9.30 Carried Load with Varied Trunk Thresholds.

cally in the overload cases just beyond the peak. Because m is high in this region, too much alternate routing is taking place and the preferred probing states are not receiving preferential treatment.

Depending on the network stress situation the optimal value of m varies. An interesting feature of these curves is that if m is set close to n, a severe drop in carried load results, which suggests that an upper limit m should be defined (in our case, perhaps m should be set at about 108 trunks). Some of the curves (for instance, the 100% overload curve) do not show any clear optimum because the traffic is so high that the network runs efficiently only by using the preferred probing states.

3.4.2   Time Behavior

In the constant overload situation ($\lambda_0 = 1500$), the carried load and link blocking behavior are plotted in Fig. 9.31 and 9.32 for the setting of $m = 108$. Note how smooth the trunk reservation control is (evidenced by the link blocking behavior) compared to the parallel situation of the access

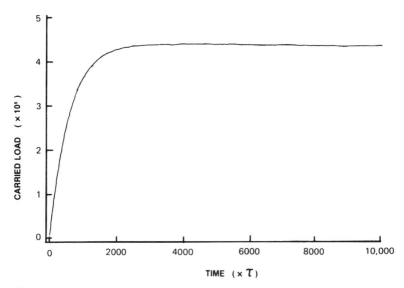

Figure 9.31   Carried Load, $m = 108$, $\lambda_0 = 1500$, Reattempts.

denial control (see Fig. 9.27). Also, the link blocking is much higher in the trunk reservation case. However, the end-to-end blocking is lower since a higher carried load is achieved using the setting for the trunk reservation threshold over the optimal setting of access denial. The reason for this is that trunk reservation discriminates against calls which use more resources in setting up a call while access denial blindly cuts back on all entering traffic. Because the "right" calls are blocked with trunk reservation, the link blocking probability appears high. If, in the access denial case, the critical blocking factor q_c were set as high as the steady-state link blocking probability with trunk reservation, the carried load would drop significantly. The lesson here is that traffic controls are most effective when they are discriminating in traffic type.

Under the stress scenario, with the setting of $m = 108$, the carried load is higher than either the uncontrolled or access denial cases. Fig. 9.32B shows the carried load in the stress situation with trunk reservation activated.

As a general assessment of trunk reservation and access denial, the trunk reservation is the preferred control.

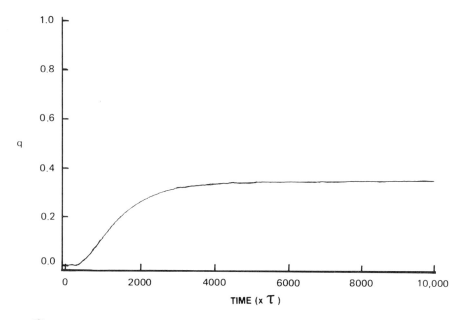

Figure 9.32 A Trunk Group Blocking, $m = 108$, $\lambda 0 = 1500$, Reattempts.

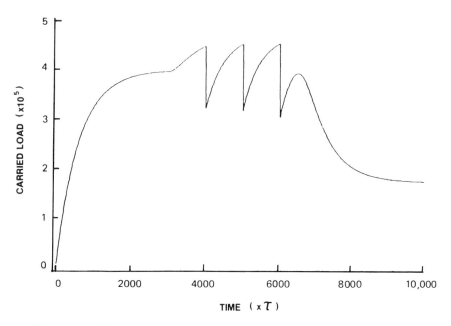

Figure 9.32B Carried Load in the Stress Scenario, $m = 108$, Reattempts.

272

3.5 Lumped Access Denial Model

We now consider access denial control for the lumped model. For this situation, the matrix A and vector B appearing in Equation (9.20) take on the form

$$
A \triangleq
\begin{bmatrix}
-\mu[p_S p_A + (1-p_A p_S)q_R] & 0 & p_R q \zeta/\tau & (p_R q \zeta/\tau) & 0 & 0 \\
p_A p_S \mu d/(N-1) & -1/\tau & 0 & 0 & 0 & 0 \\
p_A p_S \mu[1-d/(N-1)] & q/\tau & -(q\zeta+p)/\tau & (q/\tau)(1-\zeta) & 0 & 0 \\
0 & 0 & p/\tau & -1/\tau & 0 & 0 \\
0 & p/\tau & 0 & 0 & -\nu & 0 \\
0 & 0 & 0 & p/\tau & 0 & -\nu
\end{bmatrix}
\tag{9.46}
$$

where as before

$$
\zeta = \frac{p^2(1-p^2)^{r-1}}{1-(1-p^2)^r}
\tag{9.47}
$$

and

$$
B \triangleq
\begin{bmatrix}
(1-p_A p_S)p_R \\
\left(\dfrac{d}{N-1}\right) p_A p_S \\
\left(1-\dfrac{d}{N-1}\right) p_A p_S \\
0 \\
0 \\
0
\end{bmatrix}
\tag{9.48}
$$

In the above, the vector x corresponding to the state equations is as given in the lumped model section.

The state transition diagram for this lumped model is shown in Fig. 9.33, where we see incorporated the access denial probability p_A. We now must specify, as we did before, an equation describing the behavior of p_A as a function of system state. For this purpose, we choose in this section a PID (Proportional plus Integral plus Derivative) control. That is, we let

$$
q_A(t) = 1 - p_A(t) = \xi_1 \chi(t) + \xi_2 \frac{d\chi(t)}{dt} + \xi_3 \int_{t_0}^{t} \chi(t)\,dt
\tag{9.49}
$$

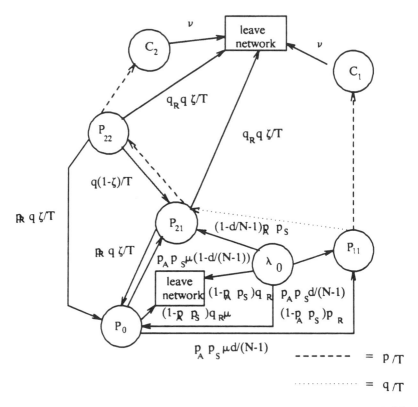

Figure 9.33   State Transition Diagram for the Lumped Access Denial Model.

where $\chi(t)$ denotes either R or q (q_B), and the parameters $\xi_i > 0$ are to be determined. The term $\xi_1 \chi(t)$ causes the control to react to the present level of the network state as measured by χ, while $\xi_2(d\chi/dt)$ fine-tunes the control by taking into account the current rate of change of the measure χ. Finally, the last (integral) term represents the memory of the control, that is, it keeps track of the variation of χ since time t_0. This last term tends to smooth out the variation in q_A; without it oscillation in q_A may occur. The critical blocking factor q_c was used for the implementation of this control. When we started from an uncongested network, the control was not applied until $q > q_c$, $q_c = 0.05$ as before, the reason for this being that some blocking (inefficiency) in the network is necessary to maximize carried load. The value of q_c used is the one resulting in a maximum carried load.

Thus, the control used, which was not applied until the first time that q became greater than q_c, is

$$p_A(t) = 1 - \xi_1(q(t) - q_c) - \xi_2\frac{dq(t)}{dt} - \xi_3\int_{t_0}^t (q(s) - q_c) \, ds$$

3.5.1 Steady-State Behavior

We begin by considering steady-state behavior using the control defined in Equation (9.49), with $\chi = q$. To find appropriate parameter settings a one-dimensional search was carried out for each parameter (ξ_1, ξ_2, ξ_3) until a change in any one of them would result in a decrease in total carried load. This search was first carried out at $\lambda = 1,500$ calls/second. The parameter values obtained were $\xi_1 = 2$, $\xi_2 = 0.5$, and ξ_3 was not used. Fig. 9.34 presents the carried load curves for the nine different damage levels as in the previous sections. As can be seen from the figure, the carried load for the top three curves begins to decrease after reaching its peak carried load; there is also noticeable instability present in the top two

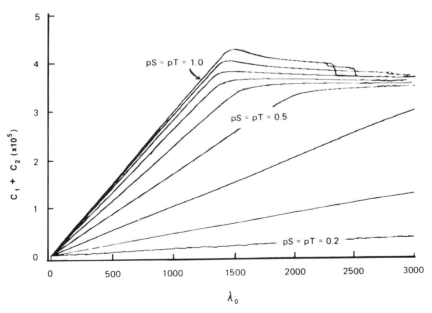

Figure 9.34 Lumped Model Carried Load.

curves. The decrease in carried load cannot be improved even when adding integral control. Note, however, that the carried load begins to decrease at around the $\lambda = 1,500$ point, the point for which the parameters were optimized. In fact, if the parameters had been optimized for $\lambda = 2,500$, the resulting curves would show that, once the carried load reaches its peak, it remains there until $\lambda = 2,500$, after which it begins to drop off. The parameters are therefore a function of the maximum offered load expected. The parameter settings for $\lambda = 3,000$ calls/second are $\xi_1 = 0.3$, $\xi_2 = 0.1$, and $\xi_3 = 0.01$. The corresponding carried load curves are shown in Fig. 9.35. Comparing these curves with the corresponding curves for the no controls case, we can see that the bistable region has disappeared. Furthermore, for the moderate damage case, it can be seen that the carried load has increased significantly. The carried load for the higher damage levels did not increase since at these levels of damage the controls are of little use. Note that as the offered load continues to increase, the carried load remains virtually constant. Fig. 9.36 and

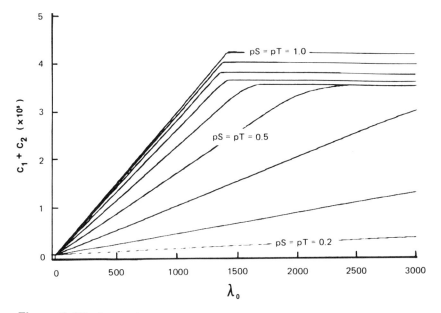

Figure 9.35 Lumped Model Carried Load.

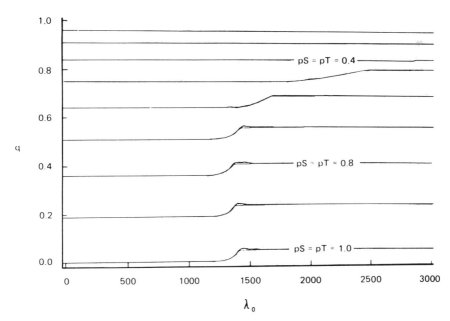

Figure 9.36 Lumped Model Blocking Probability.

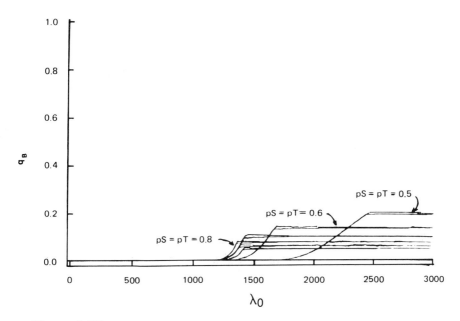

Figure 9.37 Lumped Model Blocking Probability.

9.37, show the two blocking probabilities q and q_B, respectively. Comparing these with the corresponding Fig. 9.19 and 9.20, for the no controls case, we can see that the added control is able to decrease the amount of blocking in the network caused by increased offered load. The average number of links curves for the access denial case appear in Fig. 9.38. Fig. 9.39 presents the curves for the access denial probability p_A. As can be seen the control applied to the network varies smoothly as a function of the carried load, this is a desirable feature.

3.5.2 Time Behavior

Fig. 9.40 shows a comparison in carried load between a network with no access denial control and one that incorporates access denial. Both networks were offered a constant traffic load of 1,500 calls/second, with $p_S = p_T = 1.0$. As can be seen from the figure, access denial does help in this situation of constant overload. The corresponding curves for the blocking probability q are shown in Fig. 9.41. Note that the blocking

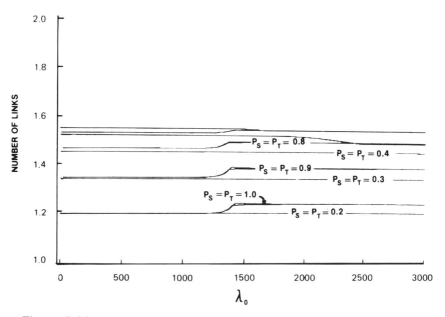

Figure 9.38 Lumped Model Average Number of Links.

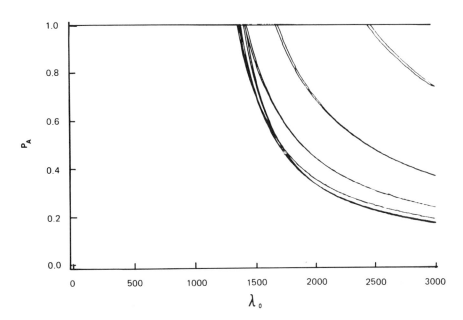

Figure 9.39 Lumped Model Access Probability.

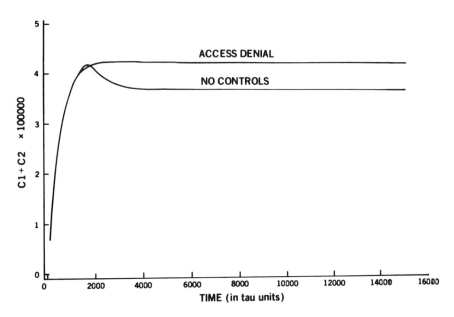

Figure 9.40 Lumped Model Carried Load.

279

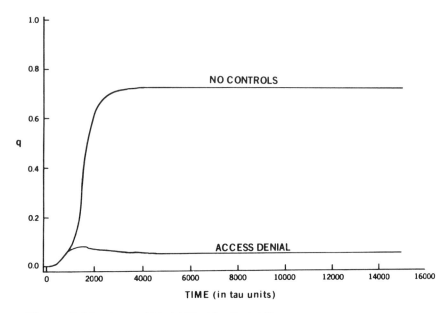

Figure 9.41   Lumped Model Blocking Probability.

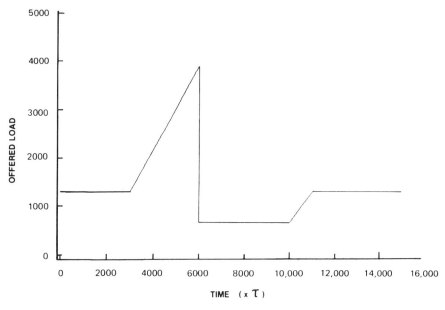

Figure 9.42   Lumped Model Offered Load.

280

probability for the access denial case is maintained at a much lower level than that of the no controls case, leading to the greater network call-carrying efficiency.

We next consider a time-varying offered load shown in Fig. 9.42. The characteristics of this traffic pattern are similar to the one previously used. The main difference, between the situation considered here and that of the previous section, is that as soon as we reach the highest offered load (approximately three times the engineered load), we quickly drop down to one-half the engineered load and after a while begin to build back up to the engineered load. Furthermore, here we do not consider damage. The corresponding carried load curves for the case of access denial control and no controls are shown in Fig. 9.43. Note that access denial control shows no improvement during the first part of the curve, simply because the network has not yet reached saturation. As soon as the traffic offered starts to surpass the engineered load, we see that access denial increases the observed carried load. However, when the traffic is suddenly cut back to one-half the engineered load, the carried load for the access denial case is

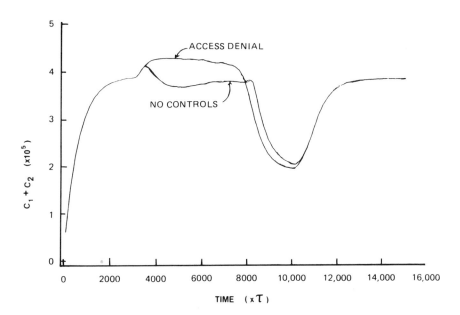

Figure 9.43 Lumped Model Carried Load.

actually lower than for the no controls case. If we look at the corresponding curves for the blocking probability q, Fig. 9.44, it can be seen that the
blocking probability for the no controls case is much higher than that of
the controls case. It can also be seen that when the traffic is suddenly cut
back to 6,000 τ, the value of the controlled q also drops rapidly; this is a
result of the derivative portion of the control responding to a sudden
change in offered load. The blocking probability q then begins to slowly
increase due to the effects of the control's integral (memory) component
and the effect of the large retrial population existing before the drop in
offered calls. It takes some time, until around $\lambda = 8,000$, for both the
controlled and uncontrolled blocking probabilities to come down to the
appropriate level for the offered load, that is, to reduce the population of
calls re-trying.

4. NETWORK OPTIMAL CONTROL

In the previous sections we have defined network models composed of state
dynamics and forcing term, or control. We have also defined various forc-

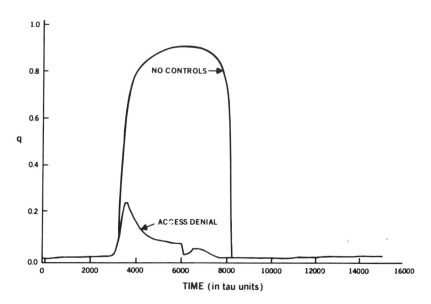

Figure 9.44   Lumped Model Blocking Probability.

ing functions and used them to control network behavior. Each forcing function used actually represented a family of functions parametrized by a set of variables, for example, ξ_i in Section 3.5. Values for these variables were selected to arrive at a "good" control. When such a choice of variables exists, it is natural to ask which choice of values for these variables results in the "best" control. This problem can be solved using optimal control techniques [3,8]. For such problems, "best" is defined through a cost functional $J(x(0),t_0,u)$ which often times takes on the form

$$J = \int_{t_0}^{t_f} L(x(\tau),u(\tau),\tau) \, d\tau + K(x_f,t_f) \qquad (9.50)$$

where t_0 denotes the initial time, t_f the final time, and x_f the final state of the system. $J(x(0),t_0,u)$ represents the cost of using the control scheme u starting with the system in state $x(0)$ at time t_0. In Equation (9.50), the function K represents the terminal cost, and the function L measures the cost of implementing the control at each instant of time. Thus if by *best* we mean that the control transfers the system from some initial state x_0 to some terminal state in minimum time, we then set $L = 1$ and $K = 0$ in Equation (9.50) and let t_f be the first instant at which the given terminal state is reached, *minimum time control*. If *best* implies that the system is to be transferred from its initial state as close as possible to a desired final state, *terminal control*, then the appropriate functional is obtained from (9.50) by letting $L = 0$ and $K = (x_f - x_d)^T K (x_f - x_d)$ where x_d denotes the desired terminal state and K is a positive definite matrix of appropriate dimension. A *minimum effort control* is obtained by letting $L = u^T R u$ and $K = 0$. Here, the best control is one that minimizes total control effort. If we have a *tracking problem* where we want to track a desired system state, $x_d(t)$, on the interval $t_0 < t < t_f$, we then let $L = [x(t) - x_d(t)]^T Q[x(t) - x_d(t)]$ and $K=0$, where Q is some positive semidefinite matrix. Of course if more than one of these objectives is to be met, then a combination of the above functionals may be appropriate [4].

If we let U denote the set of admissible controls, then our optimal control problem can be stated as follows: *Given the dynamical system*

$$\dot{x}(t) = f(x(t),u(t),t) \qquad (9.51)$$

with initial condition $x(0)$ and terminal state x_f, and initial and terminal times t_0 and t_f, respectively, find the admissible control $u \in U$ taking the system from state $x(0)$ at time t_0 to state x_f at time t_f and minimizing $J(x(0),t_0,u)$.

In the previous sections we considered two basic types of control mechanisms. The first control consists of reducing the amount of incoming traffic to the network. The other control consists of giving preferential treatment to first routed calls. We now briefly discuss the first of these as it relates to the optimal control problem.

The mechanism for reducing incoming calls was implemented through network access denial; therefore, in terms of the models presented we can consider our control to be of the form

$$u(t) = p_A \lambda_0(t) \tag{9.52}$$

If we assume that we have no control over the number of calls generated $\lambda_0(t)$, then the influx of calls into the network must be controlled through the action of p_A. In our model p_A represents the probability of being granted access into the network, as such $0 \leqslant p_A \leqslant 1$. Thus, if $\lambda_0(t)$ is greater than the needed call arrival rate, then we can always choose an appropriate p_A, that is, $p_A = \lambda_d/\lambda_0$ where λ_d is the desired call arrival rate. If $\lambda_d > \lambda_0$, then we must set $p_A = 1$, i.e., we assume that we cannot *manufacture* calls. In any event, the dynamical system that we consider in this section is that given by Equation (9.20a), namely,

$$\dot{x}(t) = A(x,t) x(t) + B(t) u(t) \tag{9.53}$$

Our goal in selecting the appropriate cost functional is to minimize probing load and control effort. By minimizing probing load we free up facilities, thereby increasing carried load. By minimizing the control effort we in essence reduce the magnitude of fluctuations in p_A, resulting in a smoother control, and maximize the portion of λ_0 that is given access to the network. Thus, we end up with a combination of *tracking control* and *minimum effort control*. The resulting cost functional J, a quadratic cost functional, is therefore given by

$$J = \int_{t_0}^{t_f} [(x - x_d)^T Q(x - x_d) + q_A^T R q_A] \, d\tau \tag{9.54}$$

where $Q \geqslant 0$ and $R > 0$ can be defined as

$$Q = \begin{bmatrix} I_4 & O \\ O & O \end{bmatrix} \tag{9.55}$$

and

$$R = 1 \tag{9.56}$$

where I_4 denotes the 4×4 identity matrix, and x_d is the state of the unstressed network when operating at maximum efficiency. The matrices Q and R need not be constant; they may also have entries other than zero or one; the actual coefficient matrices must be selected from a consideration of network characteristics. Minimizing (9.54) subject to (9.53) and $0 \leqslant p_A \leqslant 1$ will tend to maximize network carried load.

The *minimum time* and *terminal controls* do not seem to be appropriate for our situation. We are here considering networks under stress (congestion and/or damaged facilities); it is therefore not clear what the terminal state should be. Changes in minimum time may not be appropriate since such changes tend to be non-smooth (e.g., bang-bang control); this could lead to network instabilities.

5. OTHER NETWORK MODELS

In this section difference equations for related network models are presented. A port model incorporates nodes which have ports, with a port being required to utilize a node. One might think of these ports as being switchable pads in a Private Branch Exchange (PBX) or some transmission compensation required for the duration of a call. The phenomenon for the port model is interesting because the carried load can effectively produce a form of access denial. The feedback time scale for this self-inhibition is on the order of the call holding time, much longer than the feedback time intervals considered to be beneficial in stress. A special service circuit-provisioning model describes the time behavior of the demand and provisioning of private line digital channels using digital cross-connect equipment [19]. The time scales for this model are on the order of months and years, whereas the time scales for the switched systems considered in this paper are on the order of seconds and minutes. The final model we consider is a trunk reservation scheme for preferred (high-priority) customers and is similar to the trunk reservation scheme for first-routed calls.

5.1 Port Model

A model can be used to model transmission compensation or link encryption, i.e., hardware is used for any call, has a node blocking characteristic that will be denoted by q_B^S. We assume no feedback controls for this model, although the controls can be inserted by inspection from previous equations. The equations are as follows:

$$\Delta P_{11} = -P_{11} + \tau\lambda_1 \tag{9.58}$$

$$\Delta P_{21}^i = -P_{21}^i + q(1-\delta_{1i})(P_{21}^{i-1} + P_{22}^{i-1})$$
$$+ \delta_{1i}(qP_{11} + \tau\lambda_2) \tag{9.59}$$

$$\Delta P_{22}^i = -P_{22}^i + pP_{21}^i \tag{9.60}$$

$$\Delta P_0 = -\tau\mu \, [p_B^S p_S + (1-p_B^S p_S) \, q_R] \, P_0$$
$$+ p_R q(P_{21}^r + P_{22}^r)$$
$$+ \tau\lambda_0(1-p_B^S p_S) \, p_R \tag{9.61}$$

$$\Delta C_1 = -\tau\nu C_1 + pP_{11} \tag{9.62}$$

$$\Delta C_2 = -\tau\nu C_2 + pP_{22} \tag{9.63}$$

where $i = 1,\ldots,r$, and the following expressions have been defined:

$$\lambda = \lambda_0 + \mu P_0 \tag{9.64}$$

$$\lambda_1 = \frac{p_B^S p_S d\lambda}{N-1} \tag{9.65}$$

$$\lambda_2 = \frac{p_B^S p_S (N - d - 1)\lambda}{N-1} \tag{9.66}$$

$$p = p_B^T p_B^S p_S p_T \tag{9.67}$$

λ is the "renormalized" call attempt rate which includes the effect of reattempts (P_0) due to network blocking. Define quantities L_e^T and L_i^T by

$$L_e^T = \frac{C_1\nu + 2C_2\nu + P_{22}\tau^{-1}}{C_1 + 2C_2 + P_{22}} \tag{9.68}$$

$$L_i^T = \frac{2p_T p_S}{Nd\tau}(P_{11} + P_{21} + P_{22}) \tag{9.69}$$

L_i^T is the rate at which free trunks are being probed, while L_e^T is the average rate at which trunks in use are freed. The ratio $E_T = L_i^T/L_e^T$ is the "effective offered load" on the average network trunk group. If we assume Poisson arrivals, the trunk group blocking is given by

$$q_B^T = 1 - p_B^T = B(n,E_T) \tag{9.70}$$

Similarly define quantities L_e^S and L_i^S by

$$L_e^S = \frac{2C_1\nu + 3C_2\nu + P_{11}\tau^{-1} + P_{21}\tau^{-1} + 2P_{22}\tau^{-1}}{2C_1 + 3C_2 + P_{11} + P_{21} + 2P_{22}} \tag{9.71}$$

$$L_i^S = \frac{p_S}{N\tau} \left[\lambda\tau + p_B^T p_T (P_{11} + P_{21} + P_{22}) \right] \qquad (9.72)$$

L_i^S is the rate at which free ports are being probed, while L_e^S is the average rate at which ports in use are freed. Assume there are n' ports per node. The ratio $E_S = L_i^S/L_e^S$ is the effective offered load on the average network node. If we assume Poisson arrivals, the node blocking is given by

$$q_B^S = 1 - p_B^S = B(n', E_S) \qquad (9.73)$$

5.2 Special Service Circuit Provisioning

The circuit-provisioning activity modeled here represents the processing of connect and disconnect orders for circuits, such as private, high-speed data lines [2,19] on DS1 (T1) facilities. Included as part of our model is the effect of incrementing or decrementing capacity of the facilities used to route circuits, on the basis of customer demand and facility planning strategies. Alternate routing of circuits over paths which have differing costs is taken into consideration, and customer "blocking" can be used to estimate lost revenues due to under-provisioning. Whereas in [18] facility underprovisioning is handled by extraordinary (expensive) augmentation measures, the surplus demand in our model is served by finding alternate routes through the facility network. Customer retrial is modeled, reflecting the influence of customers who would wait additional time to have circuits installed. The customer retrial model represents a "willingness to pay/ wait" factor which might arise when a competitor's network has serious under-provisioning or grade-of-service problems. As usual, circuits are routed over one-link or two-link paths, and the network is assumed to be spatially homogeneous. Network costs and tariff-based revenues are modeled, with the implication that the control optimization problem is to maximize profits with respect to the facility planning strategy.

Time scales are of interest here, since they are different from those in the previous examples. In the circuit-provisioning model, the circuit holding time can be long, on the order of months or years. The circuit-provisioning time can also be lengthy, perhaps two to three months. In addition, the network capacity is changing on a somewhat longer time scale, that is, a facility planning strategy is implemented over a number of years. By contrast, in the circuit-switched network model, holding time is on the order of a few minutes, call set-up time on the order of seconds (to be compared to the circuit-provisioning time), and the trunk groups and routing policy do not change. Alternate routing in the circuit-provisioning model does not happen in real time as it does in the switched network, as the circuit-provisioning time is affected mainly by administrative factors in ordering and installing a circuit. The circuit-provisioning process is more dynamic, albeit on a longer time scale, than the switched network model because the topology (capacity) is changing on the same time scale as the circuit holding time.

Because alternate routing does not occur in real time, the "probing" states that are of interest in the switched networks do not manifest themselves in this model, except in the calculation of the effective offered load on the average T1 facility. The probing states are, in a sense, bypassed in the transition from offered load to carried load, and the network blocking probability can be calculated from a steady-state formula. Global variables of interest are then C_1, C_2, and P_0. For convenience, the time increment, τ, will be set equal to the average circuit-provisioning time, which is somewhat analogous to the call set-up time in a circuit-switched model. Additional variations from previous equations include a difference equation which changes the number of trunks, n, per facility according to a facility planning strategy, and a profit function that is considerably different from the carried load revenue function in the circuit-switched models. The offered load, λ_0, varies with time to reflect the market demand.

Throughout this paper only deterministic difference equations have been considered. Traffic (and perhaps damage) should be stochastic processes, as well as the number of trunks per facility in the case of circuit provisioning. To be consistent with the rest of the paper, the difference equation for n will be deterministic, with stochastic effects reserved for future work. With these assumptions the difference equations become:

$$\Delta C_1 = -\tau \nu C_1 + p\tau \lambda d/(N-1) \tag{9.74}$$

$$\Delta C_2 = -\tau \nu C_2 \\ + \tau \lambda [1 - pd/(N-1)][1 - (1 - p^2)^r] \tag{9.75}$$

$$\Delta P_0 = -P_0 \\ + p_R \tau \lambda [1 - pd/(N-1)][1 - (1 - p^2)^r] \tag{9.76}$$

$$\Delta n = f(n) \tag{9.77}$$

where the facility transmission probability is

$$p = 1 - B(n, E) \tag{9.78}$$

with

$$E = 2\frac{\lambda}{Ndv}\left[\frac{d}{N-1} + (1 - \frac{pd}{N-1})\frac{1 - (1-p^2)^r(1 + p)}{p^2}\right] \quad (9.79)$$

and

$$\lambda = \lambda_0 + \tau^{-1}P_0 \quad (9.80)$$

The function f represents the "facility planning policy," can be positive or negative, and depends on the global variables and various threshold parameters. The present value of profits over a horizon of T periods is

$$\sum_{k=0}^{T} \rho^k C_R \left[C_1(k\tau) + C_2(k\tau)\right] - C_M \left[C_1(k\tau) + 2C_2(k\tau), n(k\tau)\right]$$

$$- C_C[f(n[k\tau])] - C_L[f(n[k\tau])] \quad (9.81)$$

ρ is a discount factor reflecting inflation, $C_R(.)$ denotes the revenue generated by the number of circuits set up, $C_M(.,.)$ represents the costs of maintaining the circuits and facilities, $C_C(.)$ is the capital costs of installing or removing facilities, and $C_L(.)$ is the labor cost of installing or removing facilities.

5.3 Trunk Reservation Model for Preferred Customers

The offered load λ_0 is split into two components λ_0^I and λ_0^{II}, and the global variable equations for the two classes of calls are identical to those previously discussed except for the definition of the trunk blocking which will be different for the different classes. The two sets of equations are coupled through the blocking for each traffic class because the traffic load in one class affects the blocking in another. If we use the variable definitions of the previous sections, the equations for Class II variables (non-bold face) are as follows:

$$\Delta P_{11} = -P_{11} + \tau\lambda_1^{II} \quad (9.82)$$

$$\Delta P_{21}^i = -P_{21}^i + q(P_{21}^{i-1} + P_{22}^{i-1})$$
$$+ \delta_{1i}(qP_{11} + \tau\lambda_2^{II}) \quad (9.83)$$

$$\Delta P_{22}^i = -P_{22}^i + pP_{21}^i \quad (9.84)$$

$$\Delta P_0 = -\tau\mu (p_S + q_S q_R)P_0$$
$$+ p_R q(P_{21}^r + P_{22}^r) + \tau\lambda_0^{II}q_S p_R \quad (9.85)$$

$$\Delta C_1 = -\tau\nu C_1 + pP_{11} \tag{9.86}$$

$$\Delta C_2 = -\tau\nu C_2 + pP_{22} \tag{9.87}$$

where

$$\lambda^{II} = \lambda_0^{II} + \mu P_0 \tag{9.88}$$

$$\lambda_1^{II} = \frac{p_S d\lambda^{II}}{N - 1} \tag{9.89}$$

$$\lambda_2^{II} = \frac{p_S(N - d - 1)\lambda^{II}}{N - 1} \tag{9.90}$$

and

$$p = p_B p_S p_T \tag{9.91}$$

The global variables equations for Class I quantities are as follows (bold-face):

$$\Delta\mathbf{P}_{11} = -\mathbf{P}_{11} + \tau\lambda_1^{I} \tag{9.92}$$

$$\Delta\mathbf{P}_{21}^{i} = -\mathbf{P}_{21}^{i} + \mathbf{q}(\mathbf{P}_{21}^{i-1} + \mathbf{P}_{22}^{i-1})$$
$$+ \delta_{1i}(\mathbf{q}\mathbf{P}_{11} + \tau\lambda_2^{I}) \tag{9.93}$$

$$\Delta\mathbf{P}_{22}^{i} = -\mathbf{P}_{22}^{i} + \mathbf{p}\mathbf{P}_{21}^{i} \tag{9.94}$$

$$\Delta\mathbf{P}_0 = -\tau\mu(p_S + q_S q_R)\mathbf{P}_0$$
$$+ p_R \mathbf{q}(\mathbf{P}_{21}^{r} + \mathbf{P}_{22}^{r}) + \tau\lambda_0^{I} q_S p_R \tag{9.95}$$

$$\Delta\mathbf{C}_1 = -\tau\nu\mathbf{C}_1 + \mathbf{p}\mathbf{P}_{11} \tag{9.96}$$

$$\Delta\mathbf{C}_2 = -\tau\nu\mathbf{C}_2 + \mathbf{p}\mathbf{P}_{22} \tag{9.97}$$

where

$$\lambda^{I} = \lambda_0^{I} + \mu\mathbf{P}_0 \tag{9.98}$$

$$\lambda_1^{I} = \frac{p_S d\lambda^{I}}{N - 1} \tag{9.99}$$

$$\lambda_2^{I} = \frac{p_S(N - d - 1)\lambda^{I}}{N - 1} \tag{9.100}$$

and

$$\mathbf{p} = \mathbf{p}_B p_S p_T \tag{9.101}$$

Let m denote the trunk reservation threshold level, that is, the number of trunks that must be busy before Class II calls will be blocked. Define L_i, $\mathbf{L}_i$ and f by

$$L_i = \frac{2p_T p_S}{N d_T}(P_{11} + P_{21} + P_{22}) \tag{9.102}$$

$$\mathbf{L}_i = \frac{2p_T p_S}{N d_T}(\mathbf{P}_{11} + \mathbf{P}_{21} + \mathbf{P}_{22}) \tag{9.103}$$

and

$$f = \frac{\mathbf{L}_i}{L_i + \mathbf{L}_i} \tag{9.104}$$

Then, defining $E = (L_i + \mathbf{L}_i)/L_e$, which is the total effective offered load (Class I plus Class II) on the average trunk group, we find from [1] that

$$p_B = X^{-1} \sum_{i=0}^{m-1} \frac{E^i}{i!} \tag{9.105}$$

and

$$\mathbf{p}_B = p_B + X^{-1} \sum_{i=m}^{n-1} \frac{E^i}{i!} f^{i-m} \tag{9.106}$$

where

$$X = \sum_{i=0}^{m} \frac{E^i}{i!} + \sum_{i=m+1}^{n} \frac{E^i}{i!} f^{i-m} \tag{9.107}$$

APPENDIX

In this appendix the basic blocking formula used throughout the chapter will be derived. Consider a network in which each trunk group has n trunks, and there are $Nd/2$ trunk groups, where d is the number of trunk groups emanating from each of N switches. The probability of a switch surviving damage will be p_S, where as the probability of trunk group survival will be p_T. The following table shows the number of trunks used and inverse holding times for those call states which use trunks.

Call state	No. trunks used	Inverse holding time
P_{22}	1	τ^{-1}
C_1	1	ν
C_2	2	ν

The average rate at which a trunk is freed is

$$L_e = \frac{C_1\nu + 2C_2\nu + P_{22}\tau^{-1}}{C_1 + 2C_2 + P_{22}} \tag{A.1}$$

The probing states P_{11}, P_{12}, P_{22} attempt to seize trunks in time τ; therefore, the average attempt rate per trunk group is

$$L_i = \frac{2p_Tp_S}{Nd\tau}(P_{11} + P_{21} + P_{22}) \tag{A.2}$$

The damage factors p_S, p_T enter in the expression for L_i because a trunk will not have an attempted seizure unless both the trunk and the following switch are functioning.

Let p_k ($k = 0, 1, \ldots, n$) represent the probability that k trunks in the average trunk group are busy. The $\{p_k\}$ are normalized such that

$$\sum_{k=0}^{n} p_k = 1 \tag{A.3}$$

Then from Fig. 9.A.1 which shows the transition rates among the $\{p_k\}$ a differential-difference equation can be written:

$$\frac{dp_k}{dt} = L_i p_{k-1} - (kL_e + L_i)p_k + (k + 1)L_e p_{k+1} \tag{A.4}$$

For simplicity, we assume that the trunk group is in steady state, and hence must assume that L_i and L_e vary quasi-statically. While this is clearly not true, it permits us to use the standard Erlang-B formula in the difference equations. The real blocking formula [the time dependent solution to Equation (A.4)] is assumed to be adequately approximated by the steady-state expression. If we set $dp_k/dt = 0$ and define the "effective offered

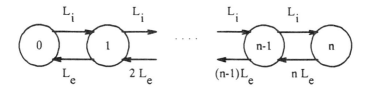

Figure 9.A.1   State Transition Graph.

load" on a trunk group, $E = L_i/L_e$, the steady-state equations for the $\{p_k\}$ become

$$Ep_{k-1} - (k + E)p_k + (k + 1)p_{k+1} = 0 \qquad (A.5)$$

The solution of Equation (A.5) subject to the normalization [Equation (A.3)] is

$$p_k = \frac{\dfrac{E^k}{k!}}{\displaystyle\sum_{i=0}^{n} \frac{E^i}{i!}} \qquad (A.6)$$

The trunk group blocking (overflow) probability q_B is the probability that n trunks are busy, i.e., p_n. Therefore,

$$q_B = p_n = B(n,E) = \frac{\dfrac{E^n}{n!}}{\displaystyle\sum_{i=0}^{n} \frac{E^i}{i!}} \qquad (A.7)$$

which is the Erlang-B blocking formula for a Poisson process of intensity E offered to n free trunks.

REFERENCES

[1] Akinpelu, J. M. - "The Overload Performance of Engineered Networks With Nonhierarchical and Hierarchical Routing, *AT&T Bell Laboratories Technical Journal* 63 (7), pp. 1261-1281, September 1984.

[2] Ashkar, G. P. Ford, G. A., Pecsvaradi, T. - "Reshaping the Network for Special Services, *AT&T Bell Laboratories Record*, pp. 4-10, September 1983.

[3] Athans, M. and Falb, P. L. - *Optimal Control*, McGraw-Hill, 1966.

[4] Barnett, S. and Cameron, R. G. - *Introduction to Mathematical Control Theory*, Clarendon Press, 1985.

[5] Burke, P. J. - "Automatic Overload Controls in a Circuit-Switched Communications Network, Proceedings of the National Electronics Conference (24), pp. 667-672, 1968.

[6] Cochrane, J. I., Falconer, W. E., Mummert V. S., Strich, W. E. - "Latest Network Trends, *IEEE Communications Magazine*, 23 (10), October 1985.

[7] Franks, R. L. and Rishel R. W. - "Overload Model of Telephone Network Operation, *Bell System Technical Journal*, 52 (9), pp.1589-1615, November 1973.

[8] Gelfand, I. M. and Fomin, S. V. - *Calculus of Variations*, Prentice Hall, 1963.

[9] Gimpelson, L. A. - "Network Management: Design and Control of Communications Networks, *Electrical Communications*, 49 (1), pp. 4-22, 1974.

[10] Girard, A. - "Blocking Probability of Noninteger Groups with Trunk Reservation, *IEEE Transactions on Communications*, COM-33 (2), pp. 113-120, February 1985.

[11] Greene, T. V., Haenschke, D. G., Hornbach, B. H., Johnson, C. E. - "Network Management and Traffic Administration, *Bell System Technical Journal*, 56 (7), pp. 1169-1202, September 1977.

[12] Haenschke, D,G., Kettler, D.A., Oberer, E. - "Network Management and Congestion in the U. S. Telecommunications Network, *IEEE Transactions on Communications*, COM-29 (4), April 1981.

[13] Jagerman, D. - "Methods in Traffic Calculations, *AT&T Bell Laboratories Technical Journal*, 63 (7), pp. 1283-1310, September 1984.

[14] Kleinrock, L. - *Queuing Systems, Volume I: Theory*, Wiley, 1975.

[15] Krupp, R. S. - "Stabilization of Alternate Routing Networks, IEEE International Communications Conference, Paper No. 31.2, Philadelphia, 1982.

[16] Mummert, V. S. - "Network Management and its Implementation on the No. 4 ESS, International Switching Symposium Record, ISS-76, pp. 241-2-1⁻-⁻241-2-7, 1976.

[17] Nakagome, Y. and Mori, H. - "Flexible Routing in the Global Communication Network, Seventh International Teletraffic Congress, Paper 426, Stockholm, 1973.

[18] Rocklin, S. M., Kashper, A., Varvaloucas, G. C. - "Capacity Expansion/ Contraction of a Facility with Demand Augmentation Dynamics, *Operations Research*, 32 (1), pp. 133-147, January-February 1984.

[19] Smith, D. R. - "A Model for Special-Service Circuit Activity, *Bell System Technical Journal* 62 (10), pp. 2911-2933, December 1983.

IV
Mixed Voice/Data Networks

10
Mixed Voice/Data Networks

Mark J. Kiemele

U.S. Air Force Academy
Colorado Springs, Colorado

1. A SIMULATION MODEL FOR INTEGRATING VOICE AND DATA

This chapter introduces, via a simulation model, the concept of an integrated voice/data network. After a brief historical sketch, the case for an integrated network is presented. Then the integrated network is addressed by considering various aspects of a simulation model.

2. HISTORICAL DEVELOPMENT

Communications and computers make an unlikely partnership in the sense that communication is as old as man and computers have just arrived on the scene. The merging of the two technologies can be traced back to around 1968. At that time computer networks took the form of time-shared systems in which sharing of CPU time was the main rationale for the network. At about the same time, the Carterfone Decision [10] gave both common carriers and noncommon carriers the legal right to expand the types of communication services that could be offered to their customers. This act, together with improved transmission technology, led to computer users sharing not only CPU time, but the resources available at the computing center as well. This included peripheral devices, memory, and software. The goal of increased resource sharing spawned numerous computer networks.

One of these is the Advanced Research Projects Agency Network (ARPANET) which is generally considered the pioneer of all computer networks. The original goals of ARPANET were to allow computer resource sharing among several Department of Defense (DOD) research centers and to provide a live research environment for exploring the technical problems involved in networking. ARPANET, once an experimental network, has expanded from a 4-node network in early 1970 to more than 250 nodes today. No longer an experimental network, ARPANET allows the sharing of distributed databases and computing resources among thousands of users at universities and research agencies across the United States and Europe.

The success of ARPANET is in good part owing to a newer method of transmission, packet switching. A packet-switched system is one in which the transmitted message is segmented into smaller fixed-size blocks or packets, each having its own copy of the destination attached. These packets traverse the network independently until they reach the destination node, where they are reassembled into the original message. ARPANET has shown that packet switching is much more efficient for the transmission of data than circuit switching. Circuit switching, on the other hand, is an older technique that was designed originally for voice transmission but has also been used for data transmission. In a circuit-switched network, a complete circuit (communications path) is established between the source and destination nodes before the start of communication, and all information flowing between these two nodes traverses the same path.

The proliferation of computer networks in the last decade is evidence that computer networking has proven itself in part as a cost-effective tool for communication activities as well as computer resource sharing. Comprehensive treatments of the classification and descriptions of existing computer-communication networks are abundant in the literature [1,2,11,12,14].

3. MOTIVATING AN INTEGRATED NETWORK

Currently military communication systems are generally designed to handle either voice calls or data transactions but not both. Such deployed systems use separate facilities for the two classes of traffic, thereby magnifying both the staff and maintenance problems that already exist. The grade of service for these systems is usually satisfactory, but crisis situations can and do force traffic flows that exceed system capabilities. Recent Defense Communication Agency (DCA) studies have shown DOD's intent to achieve an all-digital integrated network that would be operational in the late 1980s. Such a network would transmit voice and multiple classes of data (e.g., interactive, bulk, facsimile) simultaneously on a common transmission medium. The feasibility of such networks has been demonstrated by Dysart et al. [7] who contend that "the future for fully digital integrated voice and data transmission is very promising." The concept of integrating voice and data rests on the fact that speech can be digitized and thus can be handled under packet-switching schemes. No doubt one of the most significant developments of this decade is the emergence of the Integrated Services Digital Network (ISDN). Still in its infant stages, ISDN possesses the conceptual power that could transform the worldwide communication operation. The interested reader is referred to Stallings [13] for a complete treatment of both the background and details of ISDN.

Other recent studies have also addressed the problem of transmitting voice and data in the same computer-communication network. Gruber [9] aptly summarizes these studies by stating that "the motivations for considering mixed voice and data traffic . . . include: the arrival of new voice related applications with the technology now existing to economically support them, and . . . economy and flexibility. Perhaps, the best objectives of integration are . . . to realize the economics of equipment commonality, large-scale integration, higher resource utilization, and combined network operations, maintenance, and administrative policies."

The scenario to accomplish such integration has also been investigated. Despite the many tradeoffs between packet switching and circuit switching, the consensus is that circuit-switching delays have been improved to the point where both circuit switching and packet switching can be employed advantageously in the same network.

4. SWITCHING TECHNIQUES

Computer-communication networks are classified in a variety of ways. One criterion commonly used to classify networks is the communications technology that is used to transfer information from one link in a network to an adjacent link via a transmitting node. That is, the switching discipline used is generally a useful and accurate descriptor of computer networks. Early networks relied on one of two switching approaches: circuit switching or store-and-forward switching. Each of these disciplines has advantages and disadvantages; and as the communications technology improved and new ideas evolved, several other switching techniques surfaced. The following paragraphs describe the four most commonly referenced switching disciplines to date.

4.1 Circuit Switching

In a circuit-switched network, a physical path must be set up between the source node and the destination node before the start of information transfer. This path may traverse intermediate switching points (nodes). Users compete for the resources which, once acquired, remain dedicated for the duration of the transaction. Once the session is complete, the switching equipment disassembles the path and these resources are once again made available to the pool of users.

Because of its low line overhead and minimal delay, circuit switching is particularly useful in applications characterized by a steady, nonbursty flow of information. The telephone networks of the United Sates are circuit-switched systems. Users do not have dedicated voice channels but compete for limited resources. Circuit switching has remained the most effective approach for voice users, but early attempts by data users to use circuit switching resulted in problems with switching delay times and circuit utilization.

4.2 Store-and-Forward (Message) Switching

A second approach, store-and-forward or message switching, is more data-oriented. If we use this approach, a total set of information (the message) is sent from one user in the system to another by establishing a link from the "sender" to a connected node. Once the node receives the message, it must store and log it for accounting purposes until it can be forwarded to the next node in the routing scheme. Once the message is stored, the node releases the communication link. This process is repeated from node to node until the message is accepted by the "receiving" node. An acknowledgment is usually sent back through the system. Some message-switched systems break up the message into fixed blocks of information. If this is the case, then all the message blocks must be received in the prescribed order at an intermediate node before the message can be retransmitted.

While circuit switching results in a fixed message transmission delay, message switching involves variable transmission delay. Message switching with fixed routing has been used in many commercial applications (e.g., Control Data Corporation's CYBERNET). It is also the approach of DOD's AUTODIN data network. Limited adaptive routing schemes allowed store-and-forward systems to better utilize their links by allowing routing decisions to be made dynamically at intermediate nodes; however, often, link utilization was less than effective and the nodal storage requirements and message manipulation times (logging, storing, and transmitting) were sometimes excessive. As a result, an improved switching philosophy, packet switching, emerged.

4.3 Packet Switching

If we use packet switching, the system breaks a message into fixed-size "packets" (although some variable length packet schemes exist). Each packet is transmitted from source to receiver over an available link out of each intermediate node. The packets are not required to arrive at their destination in any particular order since the message reconstruction process at the receiving node is independent of the order in which the packets are received. ARPANET, for many years the mecca of networking decisions, has shown that packet switching can improve link utilization, response time, and network throughput. However, packet switching is not without its problems. Applications and research have shown that the lack of packet flow control can in some instances cause disastrous problems. As a result, many constraints have been placed on packet-switched networks. For example, trace capabilities have been added to packets and senders may be required to reserve final destination storage before shipping the message. Nevertheless, packet switching has become a dominant force in data communication systems.

4.4 Integrated Switching

The most recent approach to switching is the integrated or hybrid approach. In this technique, both circuit-switched and packet-switched components are used in the network. There are several potential scenarios for integrating voice and data in the same network. According to Dysart and others [3,5-7], one of the most promising techniques is to use circuit switching for voice traffic and packet switching for data traffic. The traffic is then integrated on a circuit-switched backbone. It is precisely this approach that has been implemented in an integrated network simulation model that we will continue to reference throughout the remainder of this book. It is important that the reader understand that it is the integrated network concepts that we wish to convey and not model implementation details. However, a simulation model such as the one used here provides

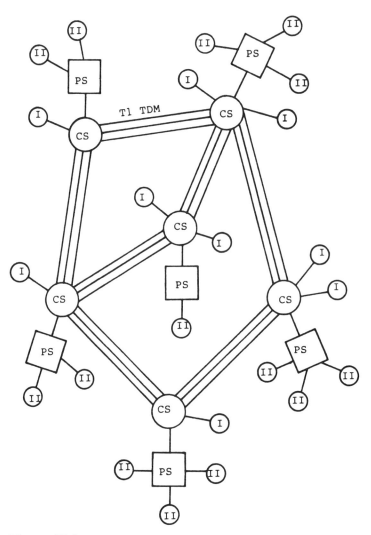

Figure 10.1   Integrated Network Configuration.

an ideal avenue through which integrated network concepts can be illustrated. The network model, or simulator, is a modified version of the network simulation model developed by Clabaugh [4]. This simulator can also be used to address strictly packet-switched networks. Using packet switching for both voice and data is currently an active area of research which shows considerable promise.

4.5   Integration Concept in the Model

The integrated network model depicts an integrated circuit/packet-switched network that consists of the following major components (Fig. 10.1):

1. Backbone Circuit-Switch (CS) Nodes
2. Peripheral Packet-Switch (PS) Nodes
3. Invariant Network Synchronous Time-Division-Multiplexed (TDM) Frame-Switching Superstructure
4. Digital Network Using T1 Carriers and Digital Switching Nodes
5. Variable Subscriber Data Rates
6. Two Classes of Subscriber Traffic
   a. Class I: Real-time traffic that once started cannot be interrupted (voice, video, facsimile, and sensor)
   b. Class II: The general class of packet data, such as interactive, bulk, and narrative/message

The backbone CS nodes and peripheral PS nodes form the nucleus of a distributed computer-communication network in which the transmission of data and voice between any two nodes on the subnet is accomplished by sharing the capacity of the T1 link. A Slotted Envelope Network (SENET) self-synchronizing concept [6] is used to achieve simultaneous transfer of voice and data on the carrier. This concept treats the available bandwidth on a digital link as a resource for which all forms of communication must compete. If we use SENET, the T1 link is synchronously clocked into frames of a fixed time duration, b, which are assumed invariant throughout the network. Each frame is partitioned into several data slots (channels) for which the various traffic types compete (Fig. 10.2). The self-synchronizing capability within each frame is implemented by using a few bits as a start-of-frame (SOF) marker to indicate the beginning of each of a contiguous series of constant period frames.

In summary, the term "integrated circuit/packet-switched network" as used here denotes a distributed computer network possessing a circuit-switched backbone or subnet with numerous packet-switched local access networks feeding into the communications subnet.

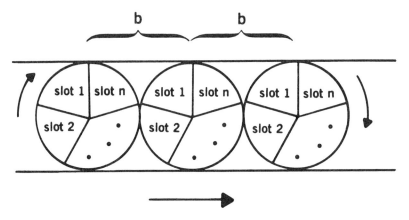

Figure 10.2   SENET Frame Clocking.

4.6   Routing Considerations

Voice (Class I) subscribers are terminated directly at circuit switches to avoid packetizing and any unnecessary routing overhead through packet switches. Similar to a telephone dial-up process, a Class I transaction results in a physical end-to-end connection for the duration of the call or a system loss (blocking) occurs. Although not a design requirement, packet switches are collocated with the circuit switches. All data (Class II) subscribers are terminated at packet switches. While the Class II subscriber packet-switch interface depends upon the individual terminal hardware configuration, the transmission of data between the packet and circuit switches is done using Time-Division Multiplexing (TDM). The packet switches are primarily responsible for managing the movement of packets between input terminals and the circuit switches; placing traffic in queues according to a regional routing policy; and performing connection initiation, circuit disconnect, and coordination with other packet switches, depending on network loading.

The regional routing doctrine for each packet switch, coupled with virtual switch connections, reduces overhead and the traffic congestion problem. As traffic is entered into a packet switch from subscriber terminals, it is queued for the relevant destination packet switch. A circuit-switch connection is then initiated/terminated by the packet switch on behalf of this traffic. A circuit-switch connection can be established on a single transac-

tion basis, similar to an interactive communication, or on a multiple transaction basis if the traffic is bulk data, message/narrative traffic, or if
several users are queued for the same destination packet switch. This routing scheme (1) insures minimal queue build-up within the backbone and (2)
enforces an end-to-end flow control strategy.

Progressive alternate routing is used on the backbone of the network.
With this method, each circuit-switch node has a primary and an alternate
path. If blocking occurs at some node during connection initiation, the
alternate route is tried for route completion. If this connection fails, the
transaction is either queued at the packet node or considered a system loss
at the circuit node, depending on its class.

4.7   Queuing Properties of the Model

For the intergrated computer-communication network described, the
numerous internodal conditions and variables preclude any exact analytic
solution. However, by decomposing the network into nodal queuing
processes, the simulation model can be viewed as a system of simple queuing models.

The traffic flow at each packet switch can be described as follows:

1. Each Class II subscriber communicates with the packet switch via an
    independent Poisson process of transaction arrivals and exponentially
    distributed transaction interarrival times.
2. The message lengths (packets per message) are assumed to be
    geometrically distributed.
3. Each packet switch can be thought of as a $M/M/C^1$ system (Kendall
    notation [8]) with infinite storage.
4. Packets are placed on the packet switch queue and served on a First-
    Come-First-Served (FCFS) basis.

The traffic flow entering each circuit-switch node originates at either neighboring circuit-switch nodes, connected packet-switch nodes, or locally terminated Class I subscribers. Since all traffic entering from other than terminated Class I subscribers sees a physical connection, only the Class I
subscribers enter into a serving mechanism process at the circuit-switch

---

[1]In this notation, the first element denotes the interarrival time distribution, the second element
denotes the service time or message length distribution, and the third element denotes the
number of servers. If a fourth element is given, it denotes the queue or buffer size available.
M stands for the (Markov) exponential distribution, G for a general or arbitrary distribution,
and D for the (deterministic) assumption of constant interarrival or service times.

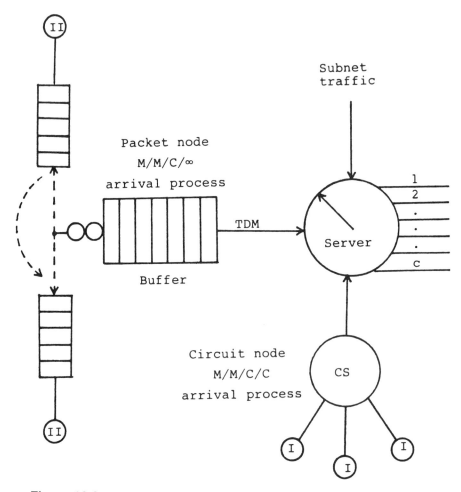

Figure 10.3 Network Nodal Queuing Processes.

node. These subscribers are assumed to possess Poisson arrival and exponential service distributions. Thus, the M/M/C/C queuing model suffices to represent this network model. Both the PS and CS queuing systems are impacted by channel availability, since the model policy forces data and voice subscribers to compete for the available slots. In summary, delays and queues at each node are approximated closely by $M/M/C/\infty$ and $M/M/C/C$ queuing processes (Fig. 10.3). Since each of these processes is globally impacted by channel availability, the network simulation provides performance measures for end-to-end packet delay and voice call blocking.

Since the communication activity is centered around the nodes in a network, the model is said to be node-based. There are three principal nodal tables: routing, channel, and queue tables. The routing tables are used to determine the output channel (link) a transaction will take from a given node in its attempt to reach its destination. Since each T1 link is full-duplex (FDX), i.e., traffic can move in both directions simultaneously, the link is represented as two independent half-duplex (HDX) channels. Information stored and updated within the various channel table entries for each node is used in the gathering of such performance statistics as link utilization and throughput. The nodal queue tables are used to obtain statistics associated with average transaction time in the system and various transaction-oriented delay and blocking statistics. From these descriptions the model is clearly capable of generating a wide variety of performance statistics.

An event table reflects network status disturbing events. Data and voice transaction arrivals or departures at a node are examples of event table entries. The simulation is driven by event changes that occur at each node.

4.8   Network Performance Measures

The network simulation model requires that the following parameters be input to the model:

1. Number of Nodes
2. Number of Links
3. Number of Time Slots for Each Link
4. Number of Slots Required by a Data Packet
5. Frame Time
6. Nodal Switching Delay
7. CS Arrival Rate
8. PS Arrival Rate
9. Simulation Start Time
10. Simulation End Time
11. PS Saturation Level Indicator
12. Voice Digitization Rate
13. PS Buffer Size

14. Voice Call Service Rate
15. Number of Bits per Packet
16. Average Number of Packets per Message

Due to the large number of system input parameters, the simulator can generate empirical data for an enormous number of topology/workload combinations. The network traffic load is determined by the voice and data arrival rates (parameters 7 and 8) and the voice call service rate (parameter 14). Link capacity is seen to be a function of parameters 3, 4, 5, and 15. The exact relationship is given by

$$Link\ capacity = \frac{1000 * PARAM(3) * PARAM(15)}{PARAM(4) * PARAM(5)}\ bits/second(bps)$$

For each possible combination of input parameters, the simulator is capable of generating more statistics than one might care to examine. The user has two statistical routines available and can have cumulative statistics printed at regular time intervals if desired. The following five performance measures that the model generates are listed here because these represent the performance measures that are most commonly seen in the literature:

1. Mean Packet Delay (MPD)
2. Average Link Utilization (ALU)
3. Packet Throughput (THR)
4. Average Queue Length (AQL)
5. Fraction of Calls Blocked (BLK)

Specification of bounds on any combination of these measures establishes what is generally called a Grade-Of-Service (GOS) for the network.

REFERENCES

[1] Abramson, N. and Kuo, F. F. (Eds.) - *Computer Communication Networks*, Prentice Hall, 1973.

[2] Ahuja, V. - *Design and Analysis of Computer Communication Networks*, McGraw-Hill, 1982.

[3] Bially, T. and McLaughlin, A. J. - "Voice Communications in Integrated Digital Voice and Data Networks," *IEEE Transactions on Communications* COM 28 (9), pp. 1478-1488, 1980.

[4] Clabaugh, C. A. - "Analysis of Flow Behavior Within an Integrated Computer-Communication Network," Ph.D. dissertation, Texas A&M University, May, 1979.

[5] Coviello, G. J. and Lyons, R. E. - "Conceptual Approaches to Switching in Future Military Networks," *IEEE Transactions on Communications* COM-28 (9), pp. 1491-1498, 1980.

[6] Coviello, G. J. and Vena, D. A. - "Integration of Circuit/Packet Switching by a SENET (Slotted Envelope Network) Concept," 1975 National Telecommunications Conference NTC-75, New Orleans, LA, pp. 42-12 - 42-17, 1975.

[7] Dysart, H., Krone, M., Fielding, J. - "Integrated Voice/Data Private Network Planning," IEEE 1981 International Conference on Communications ICC-81, Denver, CO, pp. 4.2.1-4.2.5, 1981.

[8] Gross, D. and Harris, C. M. - *Fundamentals of Queueing Theory*, Wiley, 1974.

[9] Gruber, J. G. - "Delay Related Issues in Integrated Voice and Data Networks," *IEEE Transactions on Communications* COM-29 (6), pp. 786-800, 1981.

[10] Mathison, S. L. and Walker, P. M. - "Regulatory and Economic Issues in Computer Communications," *Proceedings of the IEEE* 60 (11), pp. 1254-1272, 1972.

[11] Pooch, U. W., Greene, W. H., Moss, G. G. - *Telecommunications and Networking*, Little, Brown and Company, 1983.

[12] Sharma, R. L., de Sousa, P. T., Ingle, A. D. - *Network Systems*, Van Nostrand Reinhold Company, 1982.

[13] Stallings, W. - *Tutorial: Integrated Services Digital Networks (ISDN)*, IEEE Computer Society Press, 1985.

[14] Tanenbaum, A. S. - *Computer Networks*, Prentice-Hall, 1981.

11
Integrated Network Performance Analysis

Mark J. Kiemele

U.S. Air Force Academy
Colorado Springs, Colorado

The verification and validation of simulation models are difficult tasks. In the absence of real data from the system being simulated, the tasks seem even close to insurmountable. Action has been taken, however, to gain a high confidence level in the simulator described in Chapter 10. The motivation for this analysis stems not only from the desire to move in the direction of model verification but also from the need to obtain and analyze performance data from an integrated circuit/packet-switched computer network. Fortunately, these two encompassing goals do not diverge.

1. GOALS AND SCOPE OF THE ANALYSIS

The specific goals of the analysis are as follows:

1. Design an experiment whereby the performance data can be efficiently and economically obtained.
2. Determine the effective ranges of network parameters for which realistic and acceptable network performance results.
3. Investigate the sensitivity of performance measures to changes in the network input parameters. Specifically, determine how network performance is affected by changes in the network traffic load, trunk line or link capacity, and network size.

The scope of the analysis is restricted to those parameters that are closely related to the network topology and the workload imposed on that topology. Regarding performance sensitivity to workload and link capacity, the following four parameters are investigated:

CS: Circuit Switch Arrival Rate (voice calls/min)
PS: Packet Switch Arrival Rate (packets/sec)
SERV: Voice Call Service Time (sec)
SLOTS: Number of Time Slots per Link (a capacity indicator)

The sensitivity to network size is also investigated by varying the number of nodes and links in a network. In all cases, the performance measures observed are mean packet delay (MPD), fraction of voice calls blocked (BLK), and average link utilization (ALU).

2. DESIGN OF EXPERIMENT

The 10-node network topology shown in Fig. 11.1 was considered sufficiently complex to provide practical performance data without exhausting the computing budget.

The circuit switch (circular) nodes correspond to the major computing centers of Tymshare's TYMNET, a circuit-switched network [6]. The

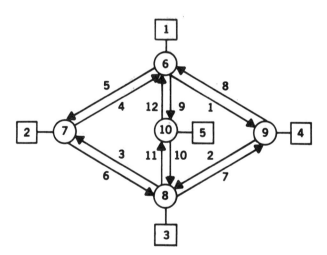

Figure 11.1   10-Node Network Topology.

trunk lines are full-duplex (FDX) carriers, each of which is modeled by two independent, half-duplex (HDX) channels.

A preliminary sizing analysis suggested that the effective range of input parameters to be investigated is as follows:

CS: 0-8 (calls/minute at each node)
PS: 0-600 (packets/sec at each node)
SERV: 60-300 (sec/call)
SLOTS: 28-52 (a link capacity indicator)

The fixed parameter settings were such that each slot represents a capacity of about 33 Kbits/sec.

The experimental design selected for this analysis is a second-order (quadratic), rotatable, central composite design [4,5]. Such a design for k (number of parameters) = 3 is illustrated in Fig. 11.2. This design was chosen because it reduces considerably the number of experimentation points that would otherwise be required if the classical 3^k factorial design were used. The "central composite" feature of the design replaces a 3^k factorial design with a 2^k factorial system augmented by a set of axial points together with one or more center points. A "rotatable" design is

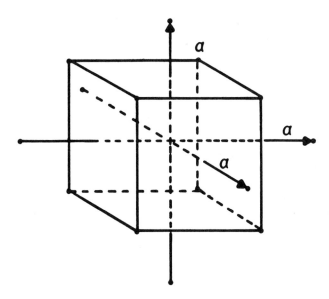

Figure 11.2   Central Composite Design ($k=3$).

one in which the prediction variance is a function only of the distance from the center of the design and not on the direction. Myers [4] shows that the number of experimental points required for $k = 4$ is 31. That is, there are 16 factorial points, eight axial points, and seven replications at the center point. In order for the design to be rotatable, α must be chosen as $(2^k)^{1/4}$. Here, α, the distance from the center point to an axial point, is 2. The seven replications at the center point will allow an estimate of the experimental (pure) error to be made; thus, a check for model adequacy is possible. Seven replications at the center point are recommended simply because it results in a near-uniform precision design. That is, it is a design for which the precision on the predicted value $\hat{y}$ given by

$$\frac{N \; var(\hat{y})}{\sigma^2}$$

where

$$N \; = \; total \; number \; of \; experimental \; points$$

$$\sigma^2 \; = \; the \; error \; variance$$

is nearly the same at a distance of one from the center point as it is at the center point [4].

In this design, each parameter is evaluated at five different levels. The five levels for each of the four parameters are as follows:

CS: 0   2   4   6   8
PS: 0   150   300   450   600
SERV: 60   120   180   240   300
SLOTS: 28   34   40   46   52

The center point is defined as CS/PS/SERV/SLOTS = 4/300/180/40.

Analysis of the results of these 31 runs suggested that the original range of data was too large. Ten of these experimental data points resulted in network performance that would be totally unacceptable, e.g., a MPD of more than 10 sec. These "outliers" were eliminated and 14 additional points in the moderate to heavy loading of the network were added. Two observations were taken at eight of these 14 additional points for estimating the error variance at points other than the center of the design. Table 11.1 shows the 43 observations that form the data set used in the subsequent regression analysis. The first 21 observations in Table 11.1 represent those data points retained from the original 31-run design. The remaining 22 observations represent the data points used to augment the original design.

Table 11.1 Experimental Data

OBS	CS	PS	SERV	SLOTS	MPD	ALU	BLK
1	2	150	120	34	0.117	0.262	0.000
2	2	150	120	46	0.116	0.194	0.000
3	2	150	240	34	0.113	0.372	0.000
4	2	150	240	46	0.116	0.275	0.000
5	2	450	120	34	0.430	0.553	0.016
6	2	450	120	46	0.260	0.396	0.000
7	2	450	240	34	0.433	0.676	0.031
8	2	450	240	46	0.449	0.481	0.003
9	6	150	120	34	0.339	0.520	0.012
10	6	150	120	46	0.115	0.386	0.000
11	6	450	120	46	0.462	0.594	0.016
12	0	300	180	40	0.123	0.231	0.000
13	4	0	180	40	0.000	0.325	0.000
14	4	300	180	52	0.153	0.427	0.000
15	4	300	180	40	0.350	0.561	0.016
16	4	300	180	40	0.527	0.571	0.005
17	4	300	180	40	0.502	0.561	0.011
18	4	300	180	40	0.316	0.581	0.020
19	4	300	180	40	0.381	0.562	0.017
20	4	300	180	40	0.798	0.608	0.036
21	4	300	180	40	0.616	0.560	0.007
22	5	400	210	43	1.613	0.739	0.070
23	5	400	210	43	2.754	0.746	0.082
24	5	400	150	37	1.656	0.725	0.067
25	5	400	150	37	1.345	0.725	0.084
26	5	400	150	43	0.192	0.615	0.027
27	5	400	210	40	4.814	0.785	0.106
28	5	400	210	40	5.624	0.782	0.125
29	4	400	210	37	1.604	0.747	0.084
30	4	400	210	37	2.730	0.745	0.101
31	4	400	210	40	2.440	0.695	0.063
32	4	400	210	40	1.199	0.687	0.057
33	4	400	180	40	0.910	0.648	0.035
34	5	400	180	40	1.970	0.722	0.074
35	5	400	180	40	1.130	0.731	0.077
36	5	400	180	37	4.084	0.775	0.110
37	5	400	180	37	4.019	0.778	0.139
38	4	400	180	37	4.817	0.712	0.057
39	4	400	180	37	1.135	0.701	0.065
40	3	400	210	37	0.599	0.652	0.017
41	3	400	210	43	0.293	0.541	0.003
42	3	400	150	37	0.454	0.570	0.003
43	3	400	150	43	0.280	0.476	0.005

3. REGRESSION ANALYSIS AND MODEL SELECTION

The Statistical Analysis System (SAS) [7-9] was used to perform a multiple regression analysis for each of the three performance measures. The regression variables are the four input parameters. SAS procedures REG and RSREG were used to analyze the data. The assumption of a quadratic response surface allows for the estimation of 15 model parameters, including the intercept.

Although the model selection procedure can be based on a number of possible criteria [1-3], the approach taken is as follows. RSREG was used first to check the full (quadratic) model for specification error (lack of fit test) and to determine significance levels for the linear, quadratic, and cross-product terms. The models for MPD, BLK, and ALU exhibited a lack of fit that was significant at the .14, .05, and .82 levels, respectively. It was found, however, that these significance levels were heavily influenced by the observations at the extremes of the heavy-loading region. For example, by eliminating 13 of the observations in the heavy-loading range of data, the lack of fit for BLK could be raised to the .70 significance level. Hence, although the .05 level for the BLK model borders on statistical significance, the lack of fit was not deemed sufficient to justify a more complex model.

The RSREG results also suggested that some terms were insignificant and possibly could be deleted from the model. With the RSREG results as a guide, several subsets of the full quadratic models were investigated using SAS procedure REG. In these models, all linear terms were retained because even though a lower-order term in a polynomial model may not be considered significant, dropping such a term could produce a misleading model [2]. Several cross-product and quadratic terms were deleted, however, and the final regression models selected are shown in Table 11.2.

Table 11.2 Regression Models

DEP VARIABLE: MPD

SOURCE	DF	SUM OF SQUARES	MEAN SQUARE	F VALUE	PROB>F
MODEL	10	61.561986	6.156199	6.874	0.0001
ERROR	32	28.658932	0.895592		
C TOTAL	42	90.220918			
ROOT MSE		0.946357	R-SQUARE	0.6823	
DEP MEAN		1.218093	ADJ R-SQ	0.5831	
C.V.		77.69169			

Table 11.2 Regression Models

VARIABLE	DF	PARAMETER ESTIMATE	STANDARD ERROR	T FOR HO: PARAMETER=0	PROB > |T|
INTERCEP	1	12.556905	13.142988	0.955	0.3465
X1 CS	1	-1.654903	1.326989	-1.247	0.2214
X2 PS	1	0.014620	0.012349	1.184	0.2452
X3 SERV	1	-0.026635	0.010541	-2.527	0.0167
X4 SLOTS	1	-0.531166	0.597025	-0.890	0.3803
X11	1	0.232675	0.078122	2.978	0.0055
X12	1	0.001153381	.0009976058	1.156	0.2562
X13	1	0.013125	0.00331975	3.954	0.0004
X14	1	-0.048214	0.025537	-1.888	0.0681
X24	1	-0.000395179	0.0003018693	-1.309	0.1998
X44	1	0.009018565	0.007302992	1.235	0.2259

DEP VARIABLE: ALU

SOURCE	DF	SUM OF SQUARES	MEAN SQUARE	F VALUE	PROB>F
MODEL	11	1.123895	0.102172	970.909	0.0001
ERROR	31	0.003262242	0.0001052336		
C TOTAL	42	1.127158			
ROOT MSE		0.010258	R-SQUARE	0.9971	
DEP MEAN		0.581233	ADJ R-SQ	0.9961	
C.V.		1.764929			

VARIABLE	DF	PARAMETER ESTIMATE	STANDARD ERROR	T FOR HO: PARAMETER=0	PROB > |T|
INTERCEP	1	0.022183	0.150510	0.147	0.8838
X1 CS	1	0.068950	0.014713	4.686	0.0001
X2 PS	1	0.001859007	0.0001334741	13.928	0.0001
X3 SERV	1	0.001788392	0.0006903726	2.590	0.0145
X4 SLOTS	1	-0.0094644	0.006917257	-1.368	0.1811
X11	1	-0.000929364	0.0008468003	-1.098	0.2809
X13	1	0.0003897799	.00003664077	10.638	0.0001
X14	1	-0.00127589	0.0002929912	-4.355	0.0001
X24	1	-.0000259887	.00000329748	-7.881	0.0001
X33	1	-.0000024518	.00000146947	-1.669	0.1053
X34	1	-.0000214946	.00000932481	-2.305	0.0280
X44	1	0.0001583866	.00008483328	1.867	0.0714

DEP VARIABLE: BLK

SOURCE	DF	SUM OF SQUARES	MEAN SQUARE	F VALUE	PROB>F
MODEL	12	0.062669	0.005222451	26.743	0.0001
ERROR	30	0.005858444	0.0001952815		
C TOTAL	42	0.068528			
ROOT MSE		0.013974	R-SQUARE	0.9145	
DEP MEAN		0.038163	ADJ R-SQ	0.8803	
C.V.		36.61764			

Table 11.2 Regression Models

VARIABLE	DF	PARAMETER ESTIMATE	STANDARD ERROR	T FOR HO: PARAMETER = 0	PROB > |T|
INTERCEP	1	0.197892	0.205845	0.961	0.3441
X1 CS	1	−0.028562	0.020039	−1.425	0.1644
X2 PS	1	0.0003439953	0.0002038975	1.687	0.1020
X3 SERV	1	0.0001307622	0.0005157113	0.254	0.8016
X4 SLOTS	1	−0.012008	0.00887909	−1.352	0.1863
X11	1	0.006553606	0.00115578	5.670	0.0001
X12	1	.00005438269	.00001502268	3.620	0.0011
X13	1	0.0003339836	0.0000496682	6.724	0.0001
X14	1	−0.00175778	0.0004020761	−4.372	0.0001
X22	1	$3.79530E-07$	$1.75917E-07$	2.157	0.0391
X24	1	−.0000158292	.00000450668	−3.512	0.0014
X34	1	−.0000194136	.00001274852	−1.523	0.1383
X44	1	0.0002843414	0.0001092323	2.603	0.0142

Despite their appearance of being insignificant, several of the cross-product and quadratic terms were retained in the model simply because their retention resulted in a smaller residual mean square than if they had been deleted. The interpretation of these regression models is presented graphically in the following section in terms of a sensitivity analysis.

4. SENSITIVITY ANALYSIS

This section presents data to describe how the various performance measures change with respect to changes in traffic load, link capacity, and network size. All the graphs presented are obtained from the quadratic response surfaces (regression models) developed in the previous section.

The workload imposed on an integrated network is described by the voice arrival rates (CS) at each circuit switch node and the data packet arrival rates (PS) at each packet switch node, as well as the length of service for each voice call (SERV). It is assumed that, for a given arrival at any particular node, all the other nodes of the same type are equally likely to be the destination node for that arrival. That is, the workload is said to be uniformly distributed between node pairs. Unless otherwise stated, it is also assumed that SERV = 180 sec. Besides these three parameters, the link capacity (SLOTS) parameter also affects the network traffic load. If

MPD(sec)

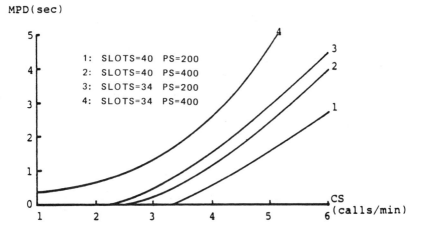

Figure 11.3 Mean Packet Delay (MPD) Versus Voice Arrival Rate (CS).

the CS and PS arrival rates are also assumed to be fixed, then the smaller values of SLOTS represent a heavier network load while the larger SLOTS values correspond to a lighter network load.

Figs. 11.3 through 11.5 depict the MPD, BLK, and ALU performance measures, respectively, as a function of CS for four different SLOTS/PS

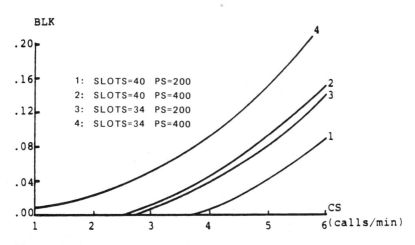

Figure 11.4 Fraction of Calls Blocked (BLK) Versus Voice Arrival Rate (CS).

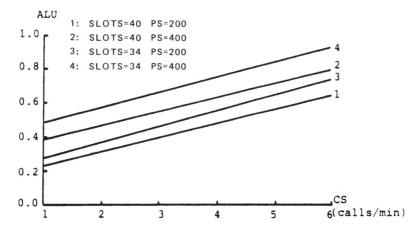

Figure 11.5   Average Link Utilization (ALU) Versus Voice Arrival Rate (CS).

combinations. Similarly, Figs. 11.6 through 11.8 show MPD, BLK, and ALU, respectively, as a function of PS for four load levels defined by combinations of SLOTS and CS. The performance measure sensitivity to the traffic load parameters CS, PS, and SLOTS (for SERV = 180) is seen by observing the relative slopes and ordinate values of the four curves in each of the Figs. 11.3 through 11.8. For example, examination of the curves in

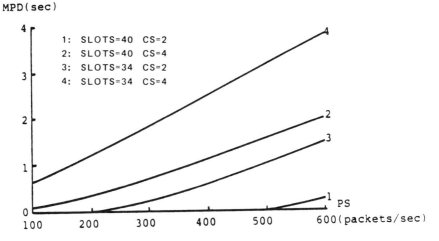

Figure 11.6   Mean Packet Delay (MPD) Versus Data Arrival Rate (PS).

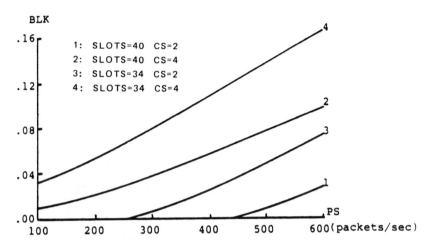

Figure 11.7   Fraction of Calls Blocked (BLK) Versus Data Arrival Rate (PS).

Figs. 11.3 and 11.6 suggests that, within the range of data shown, MPD is more sensitive to CS than to either SLOTS or PS. Additionally, the sensitivity of MPD to both CS and SERV is shown in the three-dimensional plot of Fig. 11.9, where the higher levels of one input parameter are seen to magnify the effects of the other parameter.

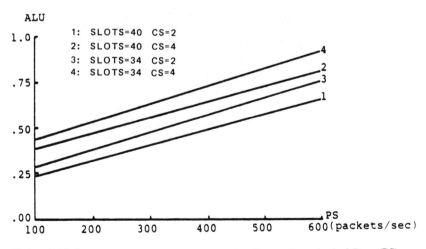

Figure 11.8   Average Link Utilization (ALU) Versus Data Arrival Rate (PS).

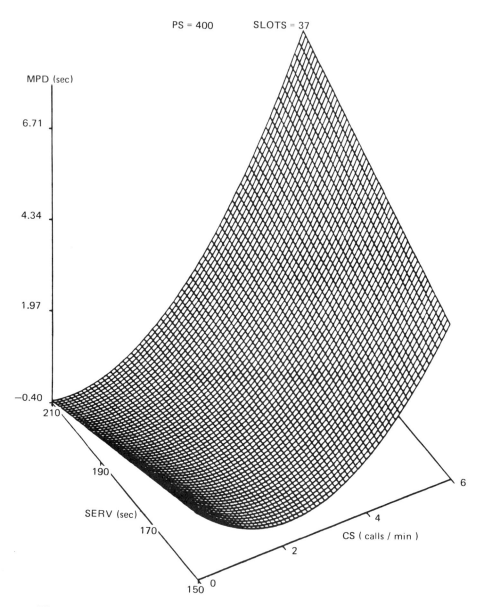

Figure 11.9 Mean Packet Delay (MPD) Versus Voice Arrival Rate (CS) Versus Voice Service Time (SERV).

The trunk line carrying capacity is an important design parameter in integrated networks. Hence, in addition to the plots of the performance measures versus number of SLOTS, which are shown in Figs. 11.10 through 11.12, the confidence intervals for each of the performance measures are also presented. Figs. 11.13 through 11.15 give the 95% confidence limits for a mean predicted value of MPD, BLK, and ALU, respectively. The voice and data arrival rates for these graphs correspond to a fairly heavy traffic load (CS = 4, PS = 400, SERV = 180).

In addition to observing network performance for a variety of traffic loads and link capacities on a fixed network topology, the analysis also examines a fixed traffic load on varying size topologiès. In particular, a throughput requirement of 2000 data packets per second with a voice call arrival rate of 20 calls per minute was imposed on three different sized networks. A 10-node, 20-node, and 52-node network were each subjected to the fixed traffic load.

The 10-node network is the TYMNET topology shown previously in Fig. 11.1. Six links interconnect the backbone nodes.

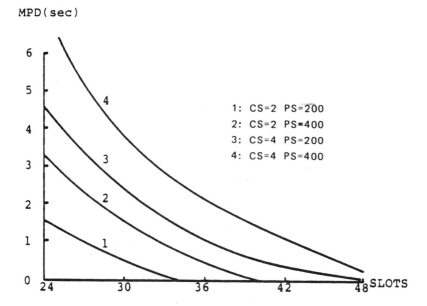

Figure 11.10   Mean Packet Delay (MPD) Versus Link Capacity (SLOTS).

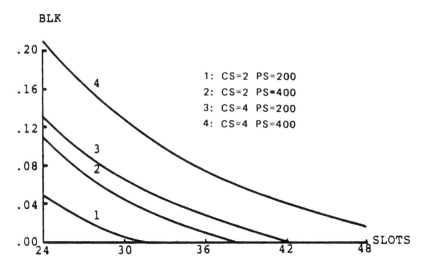

Figure 11.11   Fraction of Calls Blocked (BLK) Versus Link Capacity (SLOTS).

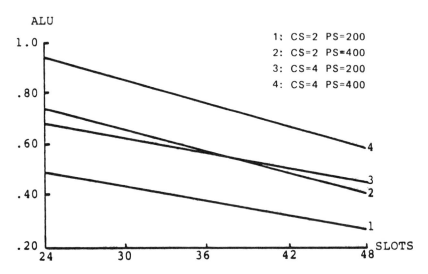

Figure 11.12   Average Link Utilization (ALU) Versus Link Capacity (SLOTS).

MPD(sec)

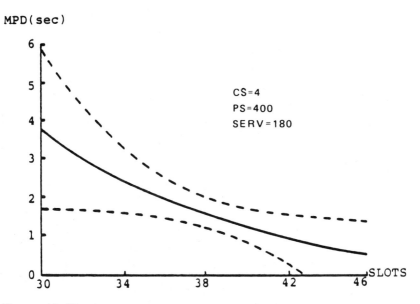

Figure 11.13   95% Confidence Limits for Mean Packet Delay (MPD).

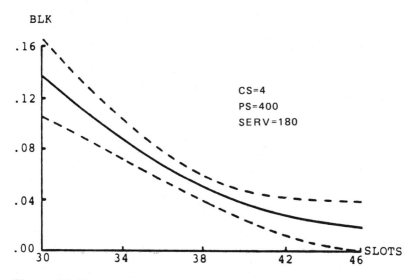

Figure 11.14   95% Confidence Limits for Fraction of Calls Blocked (BLK).

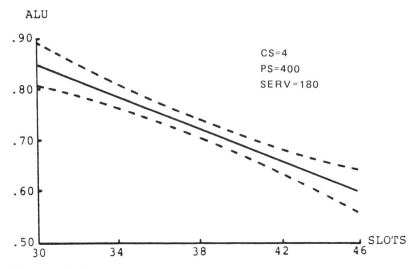

Figure 11.15 95% Confidence Limits for Average Link Utilization (ALU).

The 20-node network consists of 10 packet switches and 10 circuit switches. The 10 circuit switches forming the backbone of the network are 10 of the major computing centers in the CYBERNET network [6]. The nodes on the subnet are interconnected by 12 trunk lines. The backbone of the 20-node network is depicted in Fig. 11.16.

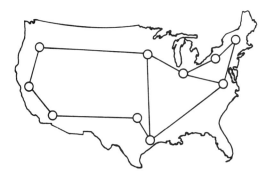

Figure 11.16 20-Node Network Backbone.

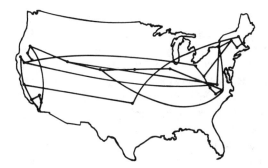

Figure 11.17   52-Node Network Subnet.

The 52-node network is comprised of 26 packet switches and 26 circuit switches, with the circuit switch nodes corresponding to a 26-node substructure of the ARPANET network. This 26-node subset of ARPANET is commonly used in the literature for comparative analyses. The subnet is interconnected with 33 links. Fig. 11.17 shows the communications subnet of the 52-node network.

All links in the three topologies have a fixed capacity of 40 slots. Additionally, each of the three communications subnets is node-biconnected (i.e., there are at least two node-disjoint paths between any pair of nodes).

Table 11.3 summarizes the results obtained when subjecting the three different network topologies to the given workload. The performance measures given for the 10-node network are averages obtained from a total of 10 simulation runs, while the performance given for the 20-node network is an average of 7 simulations. The results for the 52-node network

Table 11.3   Sensitivity to Network Size

Topology	Subnet Link/Node Ratio	MPD	BLK	ALU
10-Node	1.2	.987	.034	.654
20-Node	1.2	.451	.033	.416
52-Node	1.27	.313	.006	.289

are based on a single simulation. Although there is a reduction in MPD as the number of nodes increases, the relatively smaller decrease in MPD when going from the 20-node network to the 52-node network (as compared to going from the 10-node network to the 20-node network) results because the backbone of the integrated network is circuit-switched. This implies that there is a fixed switching delay (assumed to be 50 msec in this study) at each node, and that each packet incurs this delay at each intermediate node on its route. Hence, if the subnet link/node ratio remains fairly constant, the MPD is expected to decrease to a certain point and then begin to increase as the number of nodes in the network increases. This phenomenon is due to the fact that as more nodes are added to the network, a greater proportion of the delay can be attributed to switching delays, even though the queuing delay steadily decreases. However, as technology improves the switching delays, the effect of this phenomenon on total MPD is reduced.

Simulation model validation is a never-ending process, but a well-designed sensitivity analysis of the simulator can increase the user's confidence in the model as well as his knowledge of the model. In this regard, a sensitivity analysis is an important step in the direction of model validation. This chapter has outlined an approach to analyzing the performance characteristics of an integrated computer-communication network simulation model and has presented the results of the analysis.

The investigation has centered on the integrated network traffic load parameters of packet arrival rate (PS), voice call arrival rate (CS), and voice call service times (SERV), as well as the design parameters of link capacity (SLOTS) and network size. Network performance has been measured in terms of mean packet delay (MPD), a strict data measure; fraction of voice calls blocked (BLK), a pure voice criterion; and average link utilization (ALU), a gauge that tends to combine both the data and voice attributes of an integrated network.

The analysis has determined a range of input parameters for which second-order response surfaces can be used to adequately describe realistic network performance. This range is summarized as follows:

CS: 1-6 calls/min
PS: 100-600 packets/sec
SERV: 150-210 sec
SLOTS: 24-48 slots

Because an integrated network with a circuit-switched subnet has not yet been implemented, the term "realistic" performance is admittedly dubious. However, performance criteria for current packet-switched and circuit-switched networks can and have been used as guidelines (e.g., a packet delay of 10 sec is clearly unacceptable, for the message would automatically time out), although flexibility in the guidelines has been preserved not to stifle the range of model applicability.

Furthermore, the graphs presented in this chapter consistently support the theme that the performance measures are more sensitive to the CS (voice) arrival rate than to the other parameters investigated. The "slices" (i.e., graphs) of the response surfaces that correspond to an increased CS level generally have higher ordinate values and steeper slopes than the "slices" that correspond to increased network loads resulting from variations in the other parameters. In essence, voice arrival rates tend to dominate the network in the sense that the virtual circuits established by successful call initiations provide the framework of paths by which data packets can "piggyback" the digitized voice.

Additionally, heavily loaded networks tend to intensify the effect of any parameter. Increasing the number of nodes in a network (along with a corresponding increase in the number of links) for a fixed traffic load will decrease link utilization rates and call blocking, but may or may not decrease mean packet delay, depending on what proportion of the delay is switching delay.

The simulation model analyzed in this research is a tool that can be used by integrated network designers and managers alike. As a result of this analysis, the user of the simulation model now has a more precise understanding of the relationship between integrated network performance and the network parameters that influence such performance. In light of this, the model is a more viable tool now than it was before the analysis.

REFERENCES

[1]   Draper, N. R. and Smith, H. - *Applied Regression Analysis (2nd ed.)*, Wiley, 1981.

[2]   Freund, R. J. and Minton, P. D. - *Regression Methods*, Marcel Dekker, 1979.

[3]   Montgomery, D. C. and Peck, E. A. - *Introduction to Linear Regression Analysis*, Wiley, 1982.

[4]   Myers, R. H. - *Response Surface Methodology*, Allyn and Bacon, 1971.

[5]   Naylor, T. H. (Ed.) - *The Design of Computer Simulation Experiments*, Duke University Press, 1969.

[6]   Pooch, U. W., Greene, W. H., Moss, G. G. - *Telecommunications and Networking*, Little, Brown and Company, 1983.

[7] SAS Institute Inc. - *SAS User's Guide: Basics, 1982 Edition*, SAS Institute Inc., 1982.

[8] SAS Institute Inc. - *SAS User's Guide: Statistics, 1982 Edition*, SAS Institute Inc., 1982.

[9] SAS Institute Inc. - *SAS/GRAPH User's Guide, 1981 Edition*, SAS Institute Inc., 1981.

12

Optimizing the Topology of an Integrated Network

Mark J. Kiemele

U.S. Air Force Academy
Colorado Springs, Colorado

1. PROBLEM FORMULATION

The topology design problem for an integrated computer network can be stated as follows [2,4,10,17]:

Given: packet switch and circuit switch node locations
data traffic requirement between packet node pairs
voice traffic requirement between circuit node pairs
cost/capacity matrix
routing doctrine

Minimize: cost of the integrated network

Subject to: reliability constraint
packet delay or throughput constraint
voice call blocking constraint
link utilization constraint

Variables: link placement
link capacity

Other common formulations of the design problem are the following:

1. minimize mean packet delay given a cost constraint, and
2. maximize throughput given cost and delay constraints.

However, it has been shown [9,10] that all these formulations are closely related and that the solution techniques that apply to the stated formulation above also apply to the other formulations.

The topological design of an integrated network means assigning the links and link capacities that interconnect the circuit-switched, backbone nodes. The nodes, locations, or sites are the sources and sinks of the information flow. The data traffic matrix specifies how many packets per second must be sent between nodes i and j. Similarly, the voice traffic matrix tells how many calls per minute are initiated at node i and directed to node j. The cost/capacity matrix gives the cost per unit distance for each of the various speed transmission links available, as well as the fixed charge for each line type. There are generally only a discrete number of link capacities (speeds) available, e.g., 50 Kbits/sec, 500 Kbits/sec, 1 Mbits/sec.

The specification of constraints establishes a grade of service for the network. The reliability constraint is usually given in terms of k-connectivity. When $k=2$, a most common situation, the biconnectivity constraint suggests that there must be at least two node-disjoint paths between every pair of nodes. Mean packet delay is a common delay constraint, and it is usually given in terms of "not to exceed a specified number of milliseconds or seconds." Call blocking and link utilization are usually given in percent or a decimal fraction between 0 and 1.

2. AN ITERATIVE APPROACH TO NETWORK DESIGN

The goal of any topological design procedure is to achieve a specified performance at minimum cost. The design problem as stated above can be formulated as an integer programming problem, but the number of constraint equations quickly becomes unmanageable for even small problems. In fact, the optimal topological design solution for networks with greater than ten nodes is believed to be computationally prohibitive. This is indeed plausible since Garey and Johnson [8] have shown the network reliability problem, which is a subproblem of the topology design problem, to be NP-hard. A viable alternative to finding the optimal solution is to use a computationally efficient heuristic to generate suboptimal solutions. This is the approach illustrated here.

The technique is to generate a starting topology and evaluate this topology using the simulation model described in Chapter 10. The network topology is then perturbed in a way determined by a heuristic. The heuristic uses the performance data obtained from the simulation to move the topology in the direction of satisfying the set of constraints. The perturbed topology is once again evaluated via simulation and the heuristic reapplied. The performance feedback mechanism, or heuristic, is applied repeatedly after each simulation until all the performance constraints are satisfied, if possible. Once all the constraints are satisfied, a "feasible" solution has been obtained. The model continues to try to improve on the feasible solution by repeated use of the heuristic until it can no longer do so, at which time the best feasible solution is considered to be a "local optimum." The iterative approach is depicted in Fig. 12.1 where the heuristic is shown as a performance feedback mechanism tying the three main modules of the model in a loop.

The entire procedure can be repeated with other starting topologies, thereby generating a sequence of local optima. The situation of having many suboptimal solutions rather than the optimal solution is not all that bleak. Many factors usually enter into a design decision and modeling and analysis may be just one of them. The existence of several appropriate solutions could increase the flexibility of incorporating other nontechnical factors (e.g., political considerations) into the design decision.

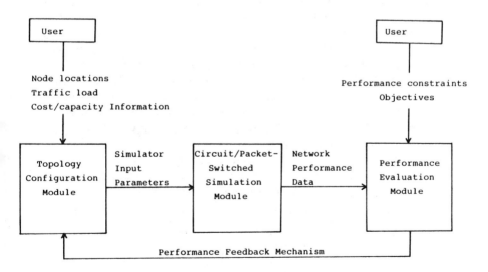

Figure 12.1   Iterative Approach.

3. STARTING TOPOLOGY GENERATION

The input to the starting topology generation process is the set of node locations, the traffic requirements, and the cost/capacity information for the set of leased lines available. The output of this process is a set of links with one of a discrete set of link capacities assigned to each existent link. That is, an integer matrix C can be used to describe a topology. If $c_{ij} = 0$, then there is no link between nodes i and j. Positive integers for c_{ij} indicate the link type or capacity between nodes i and j. Matrix C is referred to as the topology connectivity matrix and is symmetric with respect to the main diagonal, always a set of zeros. The process of generating a starting topology consists of two major steps. Each of these steps is now considered in detail.

3.1 Link Assignment

If the link assignment process were such that the resulting initial topology were feasible, many of the subsequent problems involved in heuristic selection could be avoided. The technique implemented in the model is not quite so ambitious and does not guarantee a feasible starting topology. Instead, the approach is aimed at efficiently generating a network that is reasonably close to feasibility but at the same time has a relatively low cost. The algorithm used is a modified version of a heuristic owing to Steiglitz et al. [16]. The heuristic is based on Whitney's theorem [18] which essentially states that if a network topology is k-connected, then every node in the network must have degree of at least k. The degree of a node is the number of links or arcs incident upon that node. Links in a network topology are undirected edges in a graph whose vertices correspond to the nodes in the backbone of the network. Although the condition that each node be of degree k is a necessary condition in a k-connected network, this condition is not sufficient.

The approach taken is called a link deficit approach. The difference between the required number (k) of links needed at a node and the actual degree of that node is called the link deficit for that node. The algorithm can be described as follows:

1. The nodes are randomly numbered. It is the randomization of nodes that renders the algorithm nondeterministic and that allows many starting topologies to be generated from the same input data.
2. Determine the node with the highest link deficit. Call it A. Ties are broken by the ordering of nodes.

3. Determine the node with the highest link deficit that is not already linked to node A. Call it B. Ties are broken by using minimum distance from A as a criterion or by the ordering of nodes if the distances are the same.
4. Add link AB to the network and repeat steps 2-4 until all nodes have degree of at least k.
5. If the network is connected (i.e., every node is capable of communicating with every other node), then stop.
6. Otherwise, determine the shortest link that spans two different connected components and insert this link. Go to Step 5.

Fig. 12.2 shows the result of applying steps 1-4 above to 10 nodes in the CYBERNET network (see Fig. 11.16) under the assumption that $k=2$. The numbers on each of the links suggest the order in which the links were added to the network. At this point the network is still not connected. Execution of steps 5 and 6 adds a link between nodes 3 and 6, resulting in a connected topology. Application of the algorithm guarantees a connected, but not necessarily k-connected network.

3.2 Discrete Link Capacity Assignment

The selection of link capacities from a finite set of options is the most difficult of all network design problems. Furthermore, because digitized voice and packet data are being superimposed on a common carrier, the capacity assignment problem is even more complex in an integrated network than in either the circuit-switched or packet-switched network. There are no closed form solutions even in the case where link cost is a linear function of channel capacity [12].

Complicating matters is the fact that the capacity assignment problem is intimately related to the routing problem. In order to properly assign link capacities, an estimate of the traffic load on each link is needed. But the traffic on a link is highly dependent on the routing scheme used. Extensive research has been conducted on the design and analysis of routing algorithms, and the literature abounds with routing classification schemes. One simple classification of routing schemes is found in Tanenbaum [17] who categorizes routing algorithms as either static (fixed) or dynamic (adaptive). Dynamic algorithms base their routing decisions on the current load, so consequently it is difficult to estimate traffic loads when adaptive routing is used in a network. The possibility of designing a network using one routing algorithm and operating the resulting network with another routing algorithm is real and could result in poor performance. Fortunately, it has been shown [3,9] that the performance of fixed, multiple-path routing approximates that of optimal adaptive routing under stable conditions, i.e., where the traffic load in a network is not concentrated between a small percentage of the nodes.

The approach taken to assign link capacities in an integrated network is based on the shortest distance criterion, a commonly used strategy in the literature [5]. An outline of the procedure is as follows:

1. For each pair of nodes, A and B, find the shortest path between nodes A and B and label the path as $X_1 X_2 \cdots X_n$, where $A = X_1$, $B = X_n$, and $X_2, X_3, \ldots, X_{n-1}$ are the intermediate nodes on the shortest path between A and B. Floyd's algorithm [6,15] is used to determine the shortest path between each pair of nodes.

2. For a given node pair A and B, add the packet traffic load of (A,B) or (B,A), whichever is greater, to each link on the path from A to B. Similarly, add the voice traffic load of (A,B) or (B,A), whichever is greater, to each link on the path. A voice digitization rate of 32 Kbits/sec is used to transform voice arrival rate units to link capacity units (bps).

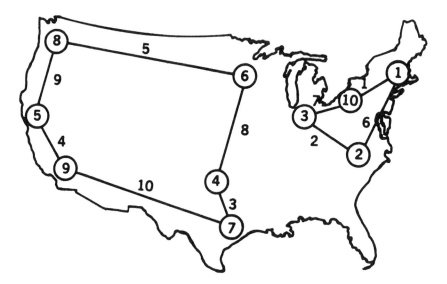

Figure 12.2 Link Deficit Approach to Assigning Links.

Table 12.1 Starting Topology Connectivity Matrix

LINE CAPACITY AND COST INFORMATION:

LINE TYPE	CAPACITY (BPS)	COST($) PER UNIT LENGTH	FIXED COST($)
1	800000.	1.00	50.00
2	1000000.	2.00	50.00
3	1200000.	4.00	100.00
4	1600000.	8.00	150.00
5	2000000.	16.00	200.00

THE RANDOMIZED NODES:

I	X(I)	Y(I)
1	25.50	13.80
2	23.10	10.90
3	19.40	10.50
4	13.40	6.10
5	1.10	11.10
6	15.10	13.90
7	14.30	4.20
8	2.80	17.20
9	2.40	8.30
10	24.30	12.40

THE CONNECT MATRIX NOW LOOKS LIKE:

	1	2	3	4	5	6	7	8	9	10
1	0	1	0	0	0	0	0	0	0	4
2	1	0	2	0	0	0	0	0	0	0
3	0	2	0	0	0	5	0	0	0	5
4	0	0	0	0	0	4	3	0	0	0
5	0	0	0	0	0	0	0	5	4	0
6	0	0	5	4	0	0	0	5	0	0
7	0	0	0	3	0	0	0	0	1	0
8	0	0	0	0	5	5	0	0	0	0
9	0	0	0	0	4	0	1	0	0	0
10	4	0	5	0	0	0	0	0	0	0

THE COST($) OF THIS TOPOLOGY IS: 2112.39

3. Repeat Step 2 for each node pair.
4. For each link in the network, assign the smallest discrete link capacity that is greater than or equal to the estimated integrated traffic load determined in Step 2. If there is no such option available, then assign the largest available link capacity.

Suppose it is assumed that each of the 10 circuit switch nodes in Fig. 12.2 has an average voice call arrival rate of 4 calls per minute and that associated with each circuit switch node is a packet switch node having a data packet arrival rate of 400 packets (1000 bits/packet) per second. Application of the above heuristic under the assumed uniformly distributed network traffic load results in the starting topology shown in Table 12.1. The topology is given as a connectivity matrix. This table also shows the cost and capacity information for the five options as well as the coordinates of the randomized nodes in the backbone. Under these conditions the cost of this starting topology is $2112.39 (a monthly charge).

4. NETWORK TOPOLOGY OPTIMIZATION

Once a starting topology has been generated and the corresponding network performance and cost ascertained, it remains to determine whether changes to the topology can enhance performance or decrease the cost or both. The procedure used to change a topology with the expectation of improving performance and/or decreasing cost stands as the crux of any iterative approach to network design. Most of the techniques seen in the literature are geared to localized transformations or minor changes that progress, it is hoped, in a stepwise fashion to a local minimum.

4.1 Perturbation Techniques

Several sources in the literature [2,4,5,7,10,11,13,14,17] provide comprehensive surveys of heuristic algorithms that have been or are being used in network design. Three of these heuristics stand out as milestone approaches to perturbing a topology, that is, changing the number of links and/or the capacity of links. These three approaches, all of which have been applied almost strictly to either packet switching or circuit switching networks, are described here to serve as a background for the approach espoused.

4.1.1 Branch Exchange Method [17]
The branch exchange (BXC) method seeks an improved design by adding links that are adjacent to deleted links. Two links are said to be adjacent if

they share a common node. Possible criteria for choosing the links to be deleted or added are link utilization rates, cost, and estimated traffic loads between node pairs. Fig. 12.3 illustrates two possible exchanges from a given starting topology. Fig. 12.3(a) is the starting network with links AD and EF assumed to be the links selected for deletion. Fig. 12.3(b) and (c) show two possible topologies that could result from the deletion of links AD and EF.

4.1.2 Concave Branch Elimination Method [9]

The concave branch elimination (CBE) method that was developed by Gerla [9] starts with a fully connected network and eliminates links until a local minimum is reached. A fully connected topology is one in which each node is connected directly to every other node. That is, if a network has N nodes, then the fully connected topology has $N(N-1)/2$ links. The scope of applicability of the CBE method is limited, however, to cases where the discrete costs can be reasonably approximated by concave functions and the packet queuing delay can be adequately described by the Pollaczek-Khintchine formula [4,5]. This formula is a concise analytical expression that gives the average queuing delay for a single-server system having Poisson arrivals and an arbitrary distribution of service times [5]. An example of link costs that can be approximated by concave functions is given in Fig. 12.4 [10].

The CBE method is computationally a more efficient design procedure than the BXC method [4], but its applicability is extremely limited. The

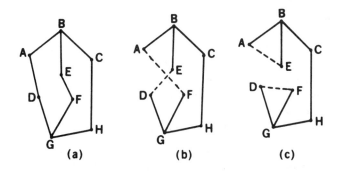

Figure 12.3 Branch Exchange Heuristic: (a) Starting Topology, (b) and (c) Resultant Topologies.

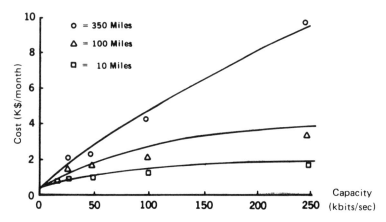

Figure 12.4 Concave Approximations to Link Costs for Various Link Lengths (copyright © 1977 IEEE reprinted by permission of IEEE).[1]

CBE heuristic is mentioned here not only because it represents a significant departure from the BXC process but also because its concept of approximating discrete costs with concave functions is used to determine bounds on heuristic performance. A complete discussion on the techniques of developing lower bounds on the cost of the optimal solution is presented in Gerla et al. [10,11]. The existence of such bounds allows the appraisal of a heuristic algorithm's accuracy.

4.1.3 Cut-Saturation Method [17]

Both the BXC and CBE methods possess inherent deficiencies. The BXC heuristic requires an exhaustive search of all local transformations and is very time-consuming when more than 20 nodes are involved. The CBE algorithm, although it can efficiently eliminate uneconomical links, does not have a link insertion capability. So once a link is deleted, there is no possibility of recovering that link. As a result of these deficiencies, a new method, the cut-saturation (CS) algorithm, evolved.

The cut-saturation method can be viewed as an improved BXC algorithm. Rather than performing all possible link exchanges as the BXC method does, the CS algorithm selects only those exchanges that are likely

---

[1] Reprinted from the paper by M. Gerla and L. Kleinrock - "On the Topological Design of Distributed Computer Networks," *IEEE Transactions on Communications* COM-25 (1), p. 57, January 1977.

to improve delay and cost. A saturated cut (or cutset) in a network is defined [10] to be the minimum set of most utilized links that, if removed, leaves the network partitioned into two disjoint components of nodes. If the links in the network shown in Fig. 12.5 are numbered in the order from most utilized link to least utilized link, then the cutset is seen to be the set $\{1, 3, 5\}$.

Tanenbaum [17] gives a scheme to find the cutset. First, rank order the links from most utilized to least utilized. Then remove from the ordered list one link at a time until the network is split into two disjoint connected components. In Fig. 12.5, this occurs after links 1-5 have been removed. To find the *minimum* cut put back each of these links into the network in turn. If putting back a link into the network does not reconnect the network, then that link is not part of the cutset. Such is the case in Fig. 12.5 for links 2 and 4. Hence, the cutset does not include these links.

The saturated cut in a network functions somewhat like a bridge between the two components. In fact, the capacity of the cutset bounds the throughput that a network could possibly realize. Hence, it seems reasonable that if a link is to be added to improve throughput or delay, that link ought to span the cutset by joining the two disjoint components. Various criteria exist as to which nodes in the two components should be connected. A commonly used criterion is to add the link having the lowest cost, which also may be the shortest link, depending on the tariff structure. Similarly, link deletions should occur only within each of the individual components. Link utilization and cost usually determine the link to be

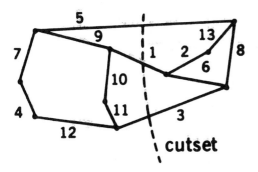

Figure 12.5   Example of a Saturated Cut.

deleted. Although minor variations of the cut-saturation algorithm exist, the substance of these algorithms remains as that described above.

A comparative analysis of the three heuristics presented here has shown that the CS algorithm gives better results and is computationally more efficient than either the BXC or CBE methods [4,10,11]. Additionally, a comparison of CS solutions to theoretical lower bounds shows that the CS algorithm can produce near-optimal solutions [11].

4.2   A Two-Phase Approach to Network Optimization

The selection of an optimization heuristic can depend on a variety of factors. Among them are the cost-capacity relationship, the topological constraints involved, the desired accuracy, the degree of human interaction, and the type of network being designed. The three perturbation techniques described above have resulted from and been applied to primarily packet-switched networks. The design and analysis of integrated or hybrid networks are in their infant stage, and the literature is almost void of any technique that is specially suited for optimizing integrated networks. Gitman et al. [12] state that there are basically two approaches to the integrated network

1. solve the link/capacity problem for the voice traffic and data traffic separately, or
2. solve the link/capacity problem for the combined voice and data traffic.

Although Gitman's approach has been to use option 1 together with a CBE heuristic [12], option 2 is the approach taken here.

A two-phase approach to the optimization of integrated networks is developed as follows. Phase 1 concerns itself with finding a feasible solution, whereas phase 2 attempts to improve performance and cost while maintaining feasibility. In phase 1, the topology is modified in a way designed to satisfy the reliability, blocking, and delay constraints. This involves using a heuristic that will in general *add* capacity to the network and consequently increase the cost. Phase 2 attempts to change the topology in a way that will decrease the cost of the network while still satisfying the three constraints. This phase involves the use of a heuristic that will in general *decrease* the total capacity of the network. Decreasing capacity means decreasing the cost and increasing the utilization, since capacity and utilization are inversely related. The entire process can be depicted as a cubic polynominal as shown in Fig. 12.6, where positive slopes are associ-

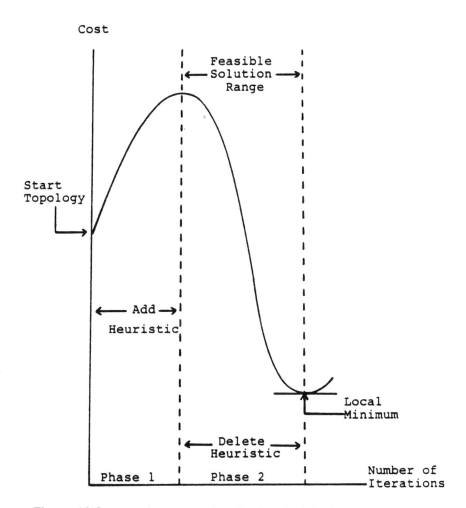

Figure 12.6 Two-Phase Approach to Topology Optimization.

ated with phase 1 and negative slopes with phase 2. A description of the heuristic used in each phase is now presented.

4.2.1   An Integrated Cut-Saturation Add Heuristic

The add heuristic used in the adaptive configuration model is based on the cut-saturation algorithm. If a link is to be added, the nodes to be connected are determined by assigning two weights to each node on the back-bone of the network. One weight is for the voice call blocking activity at the node and the other weight is for the data packet delay incurred at that node. The weights are the number of standard deviations, or z-values, that the individual node statistic (blocking or delay) deviates from the network mean statistic. If neither the delay nor blocking constraint is satisfied, then the final node weight is found by summing the two individual weights. If exactly one of these constraints is not satisfied, then only the weight of that particular statistic is taken as the final node weight. The link added is the one for which the combined weights across the cut are greater.

The add heuristic may on any one iteration add more capacity to one or more links, add one or two new links, or it may do both. An outline of the heuristic is as follows: Let

MIN = minimum average link utilization constraint,

DELTA = a decimal fraction (e.g., .10) which is an indication of the heuristic's step size,

UTIL (i) = the actual utilization of link i,

NCAPS = number of discrete capacity options,

$C(i)$ = current capacity of link i, where $C(i)$ is an integer variable that can take on the values $0, 1, 2, \ldots ,$ NCAPS,

D, B, R = indicator variables for the delay, blocking, and reliability constraints, respectively, where a 1 means the constraint is satisfied and 0 means that it is not satisfied, and

ADD = an indicator variable, where if ADD = 1, a link will have to be added (initially, ADD = 0).

1. If $D = 1$ and $B = 1$, go to Step 4.
2. For each link i such that UTIL$(i) >$ MIN, calculate $N(i) = \lceil$(UTIL(i) - MIN)/DELTA$\rceil$, where $\lceil X \rceil$ denotes the smallest integer greater than or equal to X. $N(i) \geqslant 0$ for each i. If $N(i) = 0$ for all i, then set ADD = 1.
3. For each link i, change $C(i)$ to $C(i) = C(i) + N(i)$. If $C(i) >$ NCAPS, then set $C(i) =$ NCAPS and ADD = 1.
4. If $R = 1$ and ADD = 0, then no link will have to be added, so stop.
5. If ADD = 1, then calculate the cutset and determine the components on either side of the cut.
6. If $R = 1$ and ADD = 1, then add the link determined by the weighting scheme described above and stop.
7. If $R = 0$ and ADD = 0, then a link needs to be added only to satisfy the reliability criterion. Biconnectedness ($k = 2$) is the assumed relia-

bility constraint in the model, and an algorithm to determine the biconnected components has been implemented [1]. The link chosen for inclusion is the least costly link that spans two of the biconnected components. Add the link and stop.

8. If $R = 0$ and ADD $= 1$, then at least one link is added. The link chosen is the link with highest weighting across the cut that also spans two of the biconnected components, if such a link exists. If no such link exists, then two links are added. Specifically, the links selected are the ones that would be selected in steps (6) and (7). Add the link(s) and stop.

4.2.2 A Biconnectivity-Preserving Delete Heuristic

Phase 2 of the optimization process inherits a feasible topology and seeks to reduce the cost of the network while preserving feasibility. The cost reduction, with a corresponding increase in utilization, is attained by decreasing the total capacity of the network. A biconnectivity-preserving delete heuristic is used in this approach to systematically move toward a local optimum. Topological perturbations in phase 2 are always performed on a feasible topology, and deleting a link that would reduce connectivity to something less than biconnectivity (the reliability criterion) is prohibited. The heuristic is applied to the "best" feasible topology obtained up to that point in the iterative process. This requires the storing of the "best" topology, as well as keeping a record of unsuccessful perturbations on this topology. A limit is placed on the number of consecutive perturbation failures allowed, and when this limit is reached, the "best" topology is taken as a local optimum.

The heuristic selects a link and determines whether or not the link is to be deleted. If it is not to be deleted, it then determines how much reduction in link capacity is necessary. A description of the heuristic's action is now presented. The notation is the same as in the previous section.

1. Find the least utilized link of those still qualified for investigation. Call it z.
2. If $C(z) = 1$, go to Step 5.
3. If UTIL$(z) \leq$ MIN, then calculate $N = \lceil (\text{MIN} - \text{UTIL}(z))/\text{DELTA} \rceil$. If UTIL$(z) >$ MIN, then set $N = 1$.
4. Reduce the capacity of link z by N units or to 1, whichever is greatest. That is, set $C(z) = \text{Maximum} \{C(z) - N, 1\}$. Remove z from the qualified list and stop.
5. Check to see if link z can be deleted without violating the reliability constraint. If it can, then delete z, remove z from the qualified list and

stop. If it cannot be deleted, then remove it from the qualified list and return to Step 1.

5. DESCRIPTION OF THE ADAPTIVE MODEL

The development of this adaptive topological configuration model for an integrated circuit/packet-switched computer network (CIRPAC) was based on a top-down design and a bottom-up test and integration procedure. The code was written in FORTRAN and the development and analysis of the model has been performed on an Amdahl 470.

5.1 Logical Description

A high-level logic flow of the entire design optimization process in CIR-PAC is shown in the flow diagram of Fig. 12.7. The upper loop corresponds to phase 1 and the lower loops make up phase 2 of the optimi-zation approach discussed in the previous section. To determine network feasibility, the performance characteristics of the network must be obtained. CIRPAC does this by using the simulator described in Chapter 10. The implementation of this logic in CIRPAC was accomplished by the use of six functional modules.

5.2 Modular Description

The six major modules of CIRPAC are as follows:

1. Initialization Module (INITAL)
2. Topology Configuration Module (TCM)
3. Interface Module 1 (INFACE)
4. Integrated Network Performance Generation Module (SIMULA)
5. Interface Module 2 (OUTFAC)
6. Performance Evaluation Module (PERFRM)

The names of each of these modules correspond to subroutines that are considered to be the drivers of the respective modules. Each driver sub-routine may call other subroutines that aid in accomplishing the module's function. Altogether, there are 35 routines including CIRPAC, which is considered to be the controller of the six functional modules. CIRPAC's routine calling hierarchy is shown in Fig. 12.8. A functional description of each of the six modules follows.

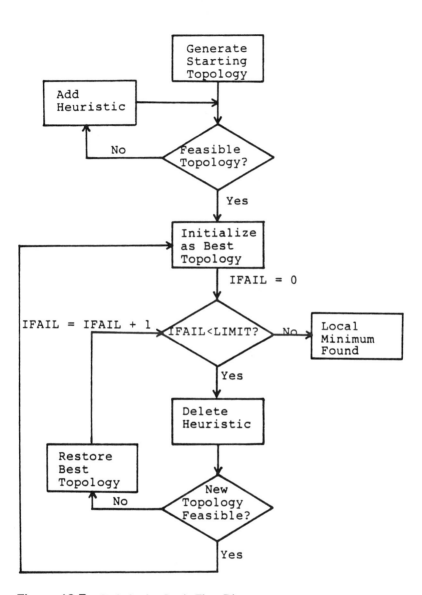

Figure 12.7   Optimization Logic Flow Diagram.

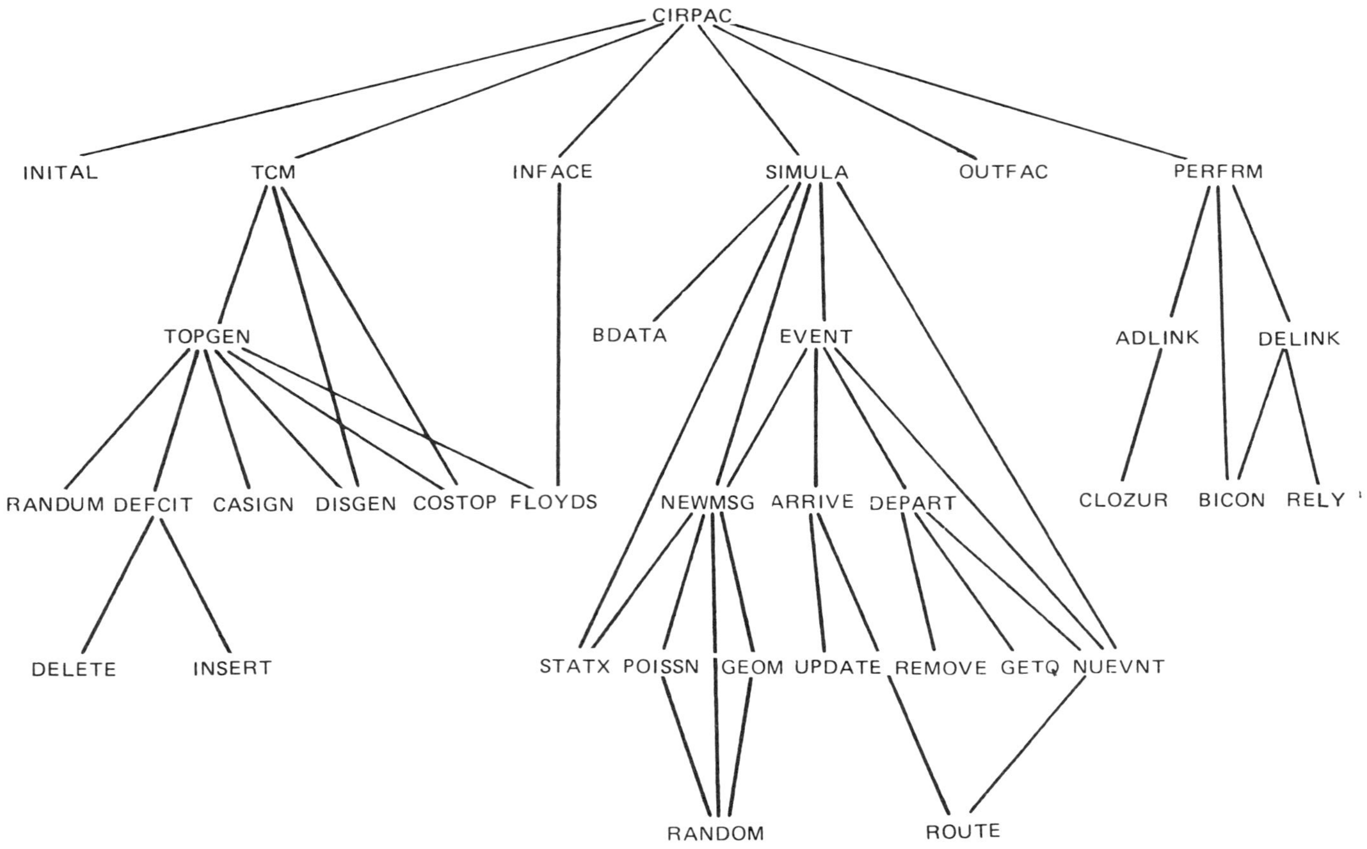

346

Figure 12.8 Routine Calling Hierarchy.

5.2.1 Initialization Module (INITAL)

The initialization module (INITAL) creates the problem domain. That is, the node or site locations, the traffic workload, the capacity/cost information, and all the constraints as given in the problem formulation are established. It also supplies the seeds for the random number generators and some control parameters that allow the user more flexibility in applying the model. For example, the user can supply a starting topology rather than having CIRPAC automatically generate one.

5.2.2 Topology Configuration Module (TCM)

The primary function of the TCM is to generate a starting topology. The TCM is capable of interpreting user-provided control parameters and directing the logic flow therefrom. Given the direction in which the topology is to move, the TCM actually modifies the topology and prints out the new topology and its associated cost at the start of each iteration. All the TCM's functions occur within the problem domain.

5.2.3 Interface Module 1 (INFACE)

The function of INFACE is to transform the current problem domain into the performance generation domain. That is, it is the interface between the TCM and the network performance generation device which, in CIRPAC, is the simulator described in Chapter 10. The purpose of distinguishing between the problem domain and the performance generation domain is to clearly divide the processes of optimizing the topology and generating network performance data. It provides for a more modularized model design. For example, if a somewhat manageable analytical procedure for providing performance data for integrated networks should appear in the future, this device could be used as the integrated network performance generation module in place of the simulator. In that case, only the interface modules would need modification to preserve the CIRPAC design tool.

INFACE is executed before each simulation of a topology, for the simulator requires its input data to be in a specific form. It initializes all seed tables and determines all simulator parameters that depend on the topology. Additionally, INFACE uses the topology data to build the routing tables used by the simulator. The primary routes are based on the shortest distance criterion, and the alternate routes are based on the minimum number of hops criteria. INFACE also checks to be sure that the dimensioning capability of the simulator is enough to handle the topology it is to simulate.

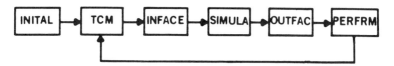

Figure 12.9 Control Flow Between Modules.

5.2.4   Integrated Network Performance Generation Module (SIMULA)

SIMULA is the simulation model described in Chapter 10, and it generates network performance data via simulation. Generating network performance of a topology means determining the performance measures of that network. SIMULA generates performance measures for voice, data, and combined voice and data. It operates totally in the performance generation domain simply because it is the performance data generator.

5.2.5   Interface Module 2 (OUTFAC)

The function of OUTFAC is to transform the network performance generation domain back into the problem domain. That is, it is the interface (on the output side of the simulator) between the network performance generation domain and the problem domain. Its function is to express the performance statistics generated in SIMULA in a form compatible with the problem domain heuristics that will use the performance data.

5.2.6   Performance Evaluation Module (PERFRM)

The performance evaluation module (PERFRM) has the function of determining how a given topology should be modified to improve performance and/or decrease cost. The heuristics that comprise the optimization technique are an integral part of this module. PERFRM has the responsibility for remembering the "best" topology as well as those perturbations that have either failed or are no longer eligible for application. PERFRM, which operates entirely in the problem domain, also decides when a local optimum has been found. This completes the descriptions of the six functional modules of CIRPAC. The flow of control (as issued by CIRPAC) between the six modules is shown in Fig. 12.9.

6.   APPLICATION OF THE ADAPTIVE MODEL

This section presents the results applying the automated methodology described as the adaptive model to several integrated networks. Integrated circuit/packet-switched networks of 20 and 52 nodes, respectively, were investigated. Unfortunately, empirical data for such integrated networks are at this time nonexistent.

6.1 20-Node Network Configuration

CIRPAC has been applied to a 20-node integrated circuit/packet-switched network to obtain 10 different local minima. The network analyzed consists of 10 backbone circuit switches and 10 peripheral packet switches where each circuit switch represents the subnet entry point for packets from exactly one packet switch. The locations of the backbone nodes are the 10 major locations of the CYBERNET and are shown as the circular nodes in Fig. 12.10. The square nodes are the collocated packet switches. The workload assumed consists of a voice call arrival rate of four calls/minute at each circuit switch and a data packet arrival rate of 400 packets/sec at each packet switch. The traffic is assumed to be symmetric and uniformly distributed between node pairs. The service time for calls is assumed to be exponentially distributed with a mean service time of 180 sec, and the capacity/cost information is as shown previously in Table 12.1.

Figure 12.10   20-Node Configuration.

When we used 10 different starting topology generator seeds, CIRPAC produced 10 different local minima for the network input data given above. The results of these 10 cases are summarized in tabular form in Table 12.2. The "start cost" is the cost of the starting topology and the "best cost" is the cost associated with the local minimum. The delay, blocking, and utilization constraints were as shown in the table. Throughput is defined to be the average number of packets/second traveling out of each node. The total number of iterations is the number of iterations needed to reach the local minimum. Biconnectivity is the assumed reliability constraint.

The local minima obtained can be used to examine integrated network design tradeoffs. For example, if each of the local minima is plotted on a utilization versus cost coordinate system, as shown in Fig. 12.11, it is seen that topology number 3 dominates all the other nine topologies. That is, each of the other minima has a higher cost *and* a lower utilization rate than case 3. The topology associated with local minimum number 3 is shown in Fig. 12.12. The integers on each link show the line type for that connection. Solid lines represent links that were part of the starting topology, while dashed lines show links that were added during the optimization process.

On the other hand, if the throughput/cost tradeoff is examined, the plot in Fig. 12.13 results. Case 3 is again dominant, but this time it dominates

Table 12.2   10 Local Minima for 20-Node Network

			AT LOCAL MINIMUM				
Case No.	Start Cost	Best Cost	≤ 1.0 sec Delay	≤ .10 Blocking	> .60 Utilization	Throughput (Packets/Sec)	No. of Iterations
1	2125.02	1925.17	.982	.083	.707	2670	15
2	2472.66	2560.29	.838	.072	.675	3000	19
3	2112.39	1882.81	.963	.085	.740	2730	14
4	2268.81	2150.61	.996	.048	.718	2756	11
5	2774.31	2160.32	.865	.080	.667	2797	14
6	2335.16	2016.40	.999	.088	.706	2639	14
7	2322.21	2386.25	.961	.080	.716	2830	14
8	2790.64	2710.77	.848	.069	.715	2837	16
9	2361.61	2175.87	.984	.083	.698	2785	16
10	2120.75	2307.08	.977	.061	.716	2811	11

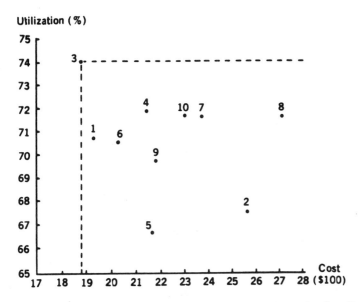

Figure 12.11   Domination of the Utilization/Cost Space by Case 3.

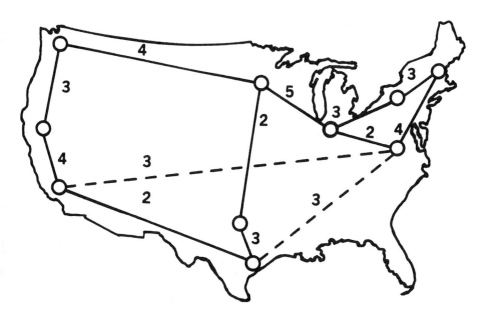

Figure 12.12   Topology for Local Minimum #3.

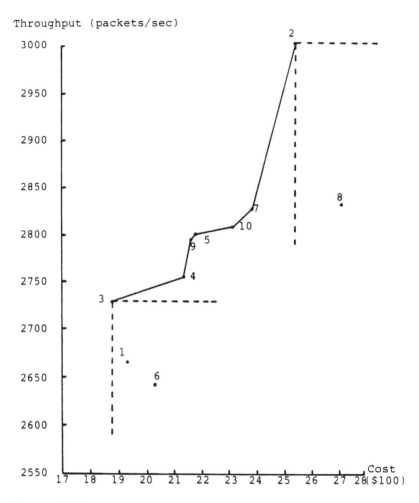

Figure 12.13   Throughput/Cost Tradeoff.

only cases 1 and 6. Additionally, case 2 dominates case 8. If all the non-dominated points are connected, the plot is capable of telling how much more throughput can be obtained for a given increase in cost. Similar investigations with other combinations of network performance measures are possible and provide the network designer with a degree of flexibility, even though the global optimum is not known.

6.2   52-Node Integrated Network

CIRPAC has also been used to optimize a 52-node integrated circuit/packet-switched network. In this case, only one local optimum has been generated. The following paragraphs describe the starting topology configurations and track the optimization process until a local minimum is reached.

The 52 nodes in the network to be optimized consist of 26 circuit switch backbone nodes and 26 peripheral packet switch nodes. As in the 20-node network, the circuit and packet switches are in a one-to-one correspondence, with each packet switch mapping onto a unique circuit switch. The 26 backbone node locations correspond to a 26-node substructure of the ARPANET. These 26 locations are commonly used in the literature when the design and analysis of packet-switched networks are addressed. Fig. 12.14 shows the backbone of the starting topology that was generated by CIRPAC. The integers on the links correspond to line type, and the underlying assumptions (e.g., workload and constraints) are the same as that for

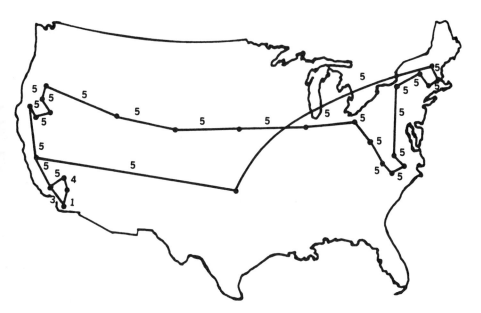

Figure 12.14   Starting Topology for 52-Node Network.

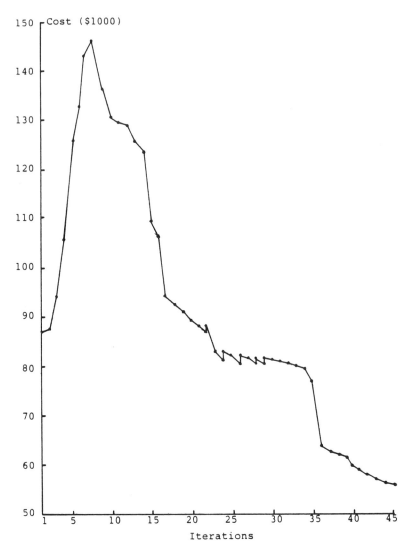

Figure 12.15   52-Node Optimization: Cost Trace.

the 20-node network discussed in the previous section. The starting topology contains 27 links and does not satisfy the biconnectivity constraint.

The process of optimizing this network with CIRPAC required a total of 45 iterations. Eight iterations were required to reach feasibility, with eight links being added in phase 1. Phase 2 required 37 iterations and two links were deleted in this phase. Fig. 12.15 shows a trace of the cost function from the starting topology to the local minimum obtained at iteration 45. The "hiccups" shown at iterations 22, 24, 26, 28, and 29 represent perturbation failures. That is, the modification to the network produced an infeasible topology, so the procedure reverted back to the "best" topology prior to the next iteration. In all five failures, the constraint failing to be satisfied was the call blocking constraint. The topology representing the local minimum is shown in Fig. 12.16. It has 33 links; 25 of the original links remain (solid lines), and 8 new links have been added (dashed lines).

In addition to the commonly used biconnectivity constraint, which is the criterion implemented in CIRPAC, several other reliability measures are also gaining wider use. Of these, two of the most common measures are

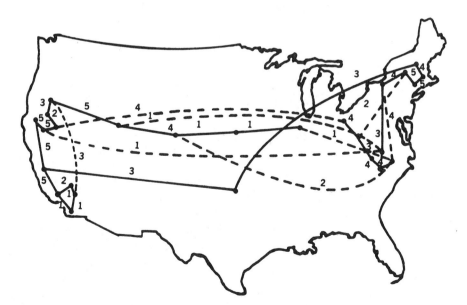

Figure 12.16 Local Minimum Topology for 52-Node Network.

the probability of the network being disconnected (PD) and the fraction of nodes unable to communicate (FR). Both of these measures are obtained by assuming that a link can fail with some nonzero probability PFAIL. If PFAIL is supplied by the user, CIRPAC will calculate both PD and FR for the local minimum obtained. This is accomplished by performing a Monte Carlo simulation on the suboptimal solution. Each simulation randomly deletes links based on the value of PFAIL, and the fraction of node pairs unable to communicate is calculated for the resulting network. Whether the resulting network is disconnected is also recorded. The number of simulations performed is based on the criterion of having a 90 percent confidence interval such that the true PD is within $\pm$ .02 of the observed PD. For the local minimum topology of the 52-node network in Fig. 12.16, PFAIL was assumed to be .05. The results of 1016 simulations on this topology are PD = .0669 and FR = .0097.

7. CONCLUDING REMARKS

This portion of the text has been directed at the concept of an integrated circuit/packet-switched computer network. It was shown that such networks are capable of carrying voice and data information simultaneously on their trunk lines. The existence of integrated networks in the near future appears to be a certainty since network designers and managers alike seek to make more efficient use of the information-carrying media available to them. The more recent innovations in the development of very high capacity transmission media, e.g., fiber optics, point to a faster transition to integrated networks than previously anticipated.

A simulation model capable of simulating an integrated network was developed to illustrate various aspects of such networks. Then a detailed network performance analysis was conducted with the simulator to ascertain the relationships between network input parameters and network performance measures. The sensitivity analysis was based on an experimental design that was especially well-suited for computer simulation experimentation. Multiple regression analyses were used to obtain models that describe the relationships between performance and network design parameters. For heavily loaded networks, the performance measures investigated were all seen to be fairly sensitive to network traffic load, link capacity, and network size. Voice call arrival rates tend to dominate the network under the progressive alternate routing strategy implemented in the simulator. An effective range of input parameters was determined for which quadratic response surfaces adequately model network performance.

The exact solution to the topological optimization problem was seen to be intractable for even small networks. The approach presented here addresses the topology design problem using an iterative, heuristic approach, whereby many suboptimal solutions (local minima) are efficiently generated in lieu of one optimal solution. The iterative scheme integrates the modified simulator as a performance generation device in the middle of a performance feedback loop. The loop consists of three processes that are repeated in turn: topology generation, network performance generation, and performance evaluation. The heuristics, part of the performance evaluation process, determine the direction in which the topology is to be modified. An integrated cut-saturation add heuristic is used to increase total network capacity until a feasible topology is obtained. Then a delete heuristic that preserves biconnectivity is used to reduce cost and increase link utilization. The methodology developed has been designed to be independent of the device that generates network performance data. Hence, if analytical means of providing adequate performance data for an integrated network should become available, only the interface modules would require change. The methodology has been applied to design problems of varying size. Results suggest that local optima can be obtained in a reasonable number of iterations even with a relatively small step size. It has been shown how the existence of several local minima can be used to actually increase the flexibility that the decision maker has in making design decisions. The model itself allows for human intervention. Starting topologies may be supplied rather than automatically generated. The number of iterations allowed is easily controlled and the stopping criteria are easily modified. Links and capacities may be forced into the topology if so desired. The methodology presented is a flexible tool that can be used to optimize the design of integrated networks.

More than 95% of the computing costs incurred in running this adaptive model can be attributed to simulation. Unless dedicated computer facilities are available for large network design, economic considerations prohibit the use of simulation as the only means for generating network performance data. Research needs to be conducted in the area of approximation techniques and analytic procedures that can provide adequate performance data for a given integrated network design specification.

With 1000-node integrated networks in the offing, the need for attacking the network design problem from a "divide and conquer" aspect is apparent. Decomposition techniques that allow for a hierarchical clustering of nodes need to be investigated. Multiple applications of a design technique on many smaller topologies could realize considerable cost savings over a single application on a large topology. A closely related area requiring attention is the design of gateways between nodal clusters that satisfy rigorous reliability constraints.

Future integrated networks will undoubtedly require priority or preemption schemes for traffic management. The distinction between data classes having different performance requirements also will most certainly be a reality. To be effective, design methodologies must be capable of addressing such issues.

The current approach of using dedicated trunk lines for voice activity needs to be reconsidered. Perhaps the high redundancy of speech can be used to develop speech buffering techniques in which minor delays may be incurred without reducing the quality of transmission. In short, voice digitization algorithms deserve increased attention.

Further investigation into the development of more effective heuristics for integrated network design is needed. The incorporation of more stringent and varied reliability criteria into the heuristic should be considered. This work has demonstrated the applicability and effectiveness of specific add and delete heuristics. Certainly other heuristics that improve the rates of convergence can be developed.

REFERENCES

[1]   Aho, A. V., Hopcroft, J. E., Ullman, J. D. - *The Design and Analysis of Computer Algorithms*, Addison-Wesley, 1974.

[2]   Ahuja, V. - *Design and Analysis of Computer Communication Networks*, McGraw-Hill, 1982.

[3]   Cantor, D. G. and Gerla, M. - "Optimal Routing in a Packet Switched Computer Network," *IEEE Transactions on Computers* C-23 (10), pp. 1062-1069, October 1974.

[4]   Chou, W. and Sapir, D. - "A Generalized Cut-Saturation Algorithm for Distributed Computer Communications Network Optimization," IEEE 1982 International Conference on Communications ICC-82, pp. 4C.2.1-4C.2.6, June 1982.

[5]   Cravis, H. - *Communications Network Analysis*, D. C. Heath and Co., 1981.

[6]   Dreyfus, S. E. - "An Appraisal of Some Shortest-Path Algorithms," *Operations Research* 17 (3), pp. 395-412, May-June 1969.

[7]   Gallager, R. - "Distributed Network Optimization Algorithms," IEEE 1979 International Conference on Communications ICC-79, pp. 43.2.1-43.2.2, June 1979.

[8]   Garey, M. R. and Johnson, D. S. - *Computers and Intractability, A Guide to*

the Theory of NP-Completeness, W. H. Freeman and Co., 1979.

[9]   Gerla, M. - "The Design of Store-And-Forward Networks for Computer Communications," Ph.D. dissertation, School of Engineering and Applied Sciences, University of California, Los Angeles, January 1973.

[10] Gerla, M. and Kleinrock, L. - "On the Topological Design of Distributed Computer Networks," *IEEE Transactions on Communications* COM-25 (1), pp. 48-60, January 1977.

[11] Gerla, M., Frank, H., Chou, W., Eckl, J. - "A Cut Saturation Algorithm for Topological Design of Packet Switched Communication Networks," 1974 National Telecommunications Conference NTC-74, San Diego, CA, pp. 1074-1079, December 1974.

[12] Gitman, I., Hsieh, W., Occhiogrosso, B. J. - "Analysis and Design of Hybrid Switching Networks," *IEEE Transactions on Communications* COM-29 (9), pp. 1290-1300, September 1981.

[13] Kleinrock, L. and Kamoun, F. - "Optimal Clustering Structures for Hierarchical Topological Design of Large Computer Networks," *Networks* 10 (3), pp. 221-248, Fall 1980.

[14] Kozicki, Z. and McGregor, P. V. - "An Approach to Computer-Aided Network Design," IEEE 1981 International Conference on Communications ICC-81, Denver, CO, pp. 4.4.1-4.4.7, June 1981.

[15] Phillips, D. T. and Garcia, A. - *Fundamentals of Network Analysis*, Prentice-Hall, 1981.

[16] Steiglitz, K., Weiner, P., Kleitman, D. J. - "The Design of Minimum-Cost Survivable Networks," *IEEE Transactions on Circuit Theory* CT-16 (4), pp. 455-460, November 1969.

[17] Tanenbaum, A. S. - *Computer Networks*, Prentice-Hall, 1981.

[18] Whitney, H. - "Congruent graphs and the connectivity of graphs," *American Journal of Mathematics* 54, pp. 150-168, 1932.

Index